D0041270

AMERICAN EVITA

ALSO BY CHRISTOPHER ANDERSEN

Sweet Caroline: Last Child of Camelot

George and Laura: Portrait of an American Marriage

Diana's Boys: William and Harry and the Mother They Loved

The Day John Died

Bill and Hillary: The Marriage

The Day Diana Died

Jackie After Jack: Portrait of the Lady

An Affair to Remember: The Remarkable Love Story of Katharine Hepburn and Spencer Tracy

Jack and Jackie: Portrait of an American Marriage

Young Kate: The Remarkable Hepburns and the Shaping of an American Legend

Citizen Jane

The Best of Everything
(with John Marion)

The Serpent's Tooth

The Book of People

Father

Susan Hayward

The Name Game

AMERICAN EVITA

Hillary Clinton's Path to Power

Christopher Andersen

wm

WILLIAM MORROW

An Imprint of HarperCollins*Publishers*

Grateful acknowledgment is made to the following for permission to reprint the photographs in this book:

AP/Wide World: 7, 13, 14, 15, 16, 17, 19, 20, 21, 22, 23, 24, 25, 26, 28, 29, 42, 44
Arkansas Democrat Gazette: 11
Jeff Christensen/SIPA: 34, 35
EPA/SIPA: 46
David P. Fornell/IPOL, Inc./Globe: 8,9
David P. Fornell/Globe: 12
Globe Photos: 4, 40
James M. Kelly/Globe Photos: 18
Rick Mackler/Rangefinders/Globe Photos: 30
Henry McGee/Globe Photos: 39
Andrea Renault/Globe Photos: 27, 31, 32, 33, 36, 43
Rickerby/SIPA: 37
Michael Schwartz/SIPA: 38
SIPA: 1, 5, 10, 41, 45, 47

AMERICAN EVITA. Copyright © 2004 by Christopher Andersen. All rights reserved. Printed in the United States of America. No part of this book may be used or reproduced in any manner whatsoever without written permission except in the case of brief quotations embodied in critical articles and reviews. For information address HarperCollins Publishers Inc., 10 East 53rd Street, New York, NY 10022.

Designed by Jo Ann Metsch

ISBN 0-06-056254-4

For my family

I don't quit.
I keep going.

—Hillary Rodham Clinton

PREFACE

She is, quite simply, the most famous, most controversial, most complex, most loved-hated-admired-reviled woman—perhaps person—in America. And, whether she fulfills her life's ambition or not, she can already lay claim to being the first woman ever considered a serious contender for the presidency.

Yet most of what we know about Hillary Rodham Clinton is strictly seen in the context of her marriage to the forty-second President. Indeed, they were the ultimate power couple—he the drawling, fatally charismatic Bubba with Falstaffian appetites, she the brilliant lawyer and consummate political strategist who put her own dreams of high office on hold to focus on capturing the Oval Office for her husband. When it came to her husband's philandering, Hillary had always been willing not just to look the other way, but to go on the offensive—as she did when news of Bill's affair with Gennifer Flowers threatened to derail Clinton's 1992 presidential campaign. "I'm not," she famously told CBS's *60 Minutes* at the time, "some Tammy Wynette, standin' by her man."

Of course, that was precisely what Hillary was, had always been, and would continue to be—particularly if it meant that she could serve, as Bill Clinton himself put it, in the role of "co-President." Hillary may have been an ardent feminist and an accomplished career woman, but the fact remained that whatever power she wielded was derived from her husband—the born political animal who, by sheer force of his formidable personality, could so effortlessly seduce voters. At the same time, it is difficult to imagine that he would ever have made it out of Arkansas were it not for having Hillary—the consummate strategic thinker, cheerleader, string puller ("Pay no attention to the woman behind the curtain!"), and excuse maker—guiding him every step of the way. "Theirs was a union of brains and sex appeal," explained one longtime ally. "Her brains and his sex appeal."

Neither her political acumen nor her impressive grasp of the weightiest issues would figure much in helping Hillary establish a political beachhead of her own. Ironically, it was in the role of the ultimate wronged woman that Hillary would shine. By the time her husband's historic impeachment trial ended with his acquittal in February of 1999, Hillary was riding an unprecedented wave of public sympathy that would later sweep her into the United States Senate—and position her for her own presidential try.

The oft-told story goes more or less like this: The Clintons are out for a drive and pull into a gas station. Hillary points to the attendant and says to her husband, "I used to date that guy." Bill laughs and says, "If you'd married him, you'd have been stuck here instead of married to the President of the United States."

"No," replies Hillary. "If I'd married him, *he'd* be President."

It is a bit of Clinton apocrypha that nevertheless has the unmistakable ring of truth. Not only did Hillary act as her husband's chief political strategist and policy sounding board, but she also played an invaluable role as lifeguard. Time and again, Hillary blew the whistle around her neck, grabbed a life preserver, and dove headfirst into the waves—rescuing her husband just as he came up for air one last time.

"She doesn't know," Clinton adviser Mandy Grunwald said at the height of the Monica Lewinsky affair, "whether to kill him or save him." In truth, there was never any real doubt. Even when Hillary seriously considered packing up and leaving her husband, their *partnership* and her faith in it remained stronger than ever.

Hillary's physical, intellectual, and emotional investment in that partnership was still paying dividends in 2004, as polls showed more members of her party favoring Senator Clinton for President than all the other declared Democratic candidates combined. Even after Massachusetts Senator John Kerry secured the Democratic presidential nomination and began to mount what appeared to be a serious challenge to incumbent George W. Bush, speculation concerning Hillary's plans for a future presidential run of her own—either in 2008 or in 2012—continued unabated.

Whenever Hillary makes her announcement, there will be the inevitable comparisons to Argentina's legendary Eva Perón—an ambitious, strong-willed woman who engineered her husband Juan Perón's rise to the presidency, used her position as Argentina's First Lady to wield enormous political power, and in the process became one of history's most admired, hated, feared, and revered women. Sex, power, money, lies, scandal, tragedy, and betrayal were the things that defined the public lives of both women. Yet more than a half century after Eva Perón's death at age thirty-three, millions regard her as nothing less than a saint—and have long lobbied the Vatican to officially make her one.

Even Hillary would concede that, in her case, sainthood seems highly unlikely. Whatever the ultimate judgment of history, the ongoing saga of Hillary Clinton's inexorable rise to power continues to stir passions—and to fan the flames of controversy that make her the *American Evita*.

Literally, I have been accused of everything from murder on down.

■

I cannot be insulted. You know? I just can't be.

My life has been kind of an unfolding drama, to me as well as everybody else.

1

The White House
Friday, January 19, 2001

Hillary Clinton was furious. Furious at the U.S. Supreme Court for handing the presidency to George W. Bush. Furious at George W. Bush for pushing his obvious advantage in Florida (where his brother was governor) to wrest control of that state's decisive electoral votes, and furious at Al Gore for blaming his defeat on the Clintons' own scandal-stained reputation.

In these waning days of their administration, the one person she was not furious at—for a change—was her husband. Throughout their marriage, it had always been Bill who screwed up and Hillary who came to the rescue. She had chosen to overlook his myriad past indiscretions as governor of Arkansas, and during their eight years in the White House stood squarely with Bill in the face of Whitewater and Travelgate and Filegate and Vince Foster and

Paula Jones and the mother of all Clinton scandals, Monicagate. Hillary, in fact, went far beyond merely standing by her man. It was the First Lady who confronted each crisis head-on, masterminding legal strategies and mounting counterattacks to debunk charges and discredit those with the audacity to have made them.

Now it was Bill's turn, and he did not have to be told what was expected of him. For years, White House staffers had been murmuring about "The Plan," the long-standing agreement that, once the Clintons left the White House, they would reverse roles: in return for all the sacrifices Hillary had made over the years—all the dreams and ambitions put on hold, not to mention the heartache and searing humiliation she had had to endure because of his rampant womanizing—Bill would throw himself behind his wife's political career. If all went according to The Plan, he would return to the White House as America's first First Gentleman. Hillary had already taken a step toward making The Plan a reality; just sixteen days earlier, she had been sworn in as the junior United States senator from New York—the only First Lady ever elected to office.

It would be hard to overstate the potential historic significance of The Plan. After all, only the first half had been implemented thus far. If all went according to schedule, Hillary would serve two terms in the White House—a combined total of sixteen years during which the Clintons would share power in the Oval Office. That would far outdistance the presidency of Franklin Roosevelt, who was elected to serve sixteen years but died after twelve. Constitutionally, there was nothing to prohibit a continuation of the informal, his-and-hers "co-presidency" the Clintons had always practiced.

In the meantime, there were some pressing issues to contend with—foremost among them the President's eleventh-hour deal to avoid prosecution in the Monica Lewinsky case. In secret meetings with independent counsel Robert W. Ray, Bill had hammered out an arrangement whereby he would admit to wrongdoing, pay a $25,000 fine, and agree to have his Arkansas law license suspended for five years. Hillary worked with her husband on the characteristically contorted wording of his so-called confession. "I tried to walk a fine line between acting lawfully and testifying falsely," he admitted, "but I now recognize that I did not fully accomplish this goal and that certain of my responses to questions about Ms. Lewinsky were false."

Neither Hillary nor Bill gave the slightest indication that, behind closed doors at the White House, they had been negotiating with Ray for weeks in a desperate effort to stave off indictment. During that time, Hillary and Bill had smiled gamely through countless farewell parties, pumped the hands of hundreds of staff members and supporters, and churned out a steady stream of heartfelt thank-you notes. Tonight, their last in the White House, they would drag themselves to one last, emotion-charged function—this one an engagement party for her longtime press aide Kelly Craighead. "He could barely stand up, he looked so tired," said a guest. "But Hillary, even though she had bags under her eyes and had been working just as hard as he had, well, she looked energized."

Hillary looked so energized, in fact, that when several aides fantasized about playing some sort of practical joke on "W" and his incoming administration, Hillary nodded her approval. "Wouldn't it be hysterical," she said with a wry smile, "if someone just happened to remove all the *w*'s from the computer keyboards?" Taking Hillary at her word, outgoing staffers dashed from office to office plucking the offending *w* keys from scores of keyboards. Others went much further, pouring coffee into file cabinets, overturning desks, leaving X-rated messages on voice-mail machines, soiling carpets, tinkering with computers, and drawing obscene pictures on office walls. (Unlike Hillary, Tipper Gore would later apologize for the vandalism of government property and the disrespect shown toward the incoming president and his family.)

While younger staffers carried out what they believed to be the First Lady's wishes, Bill, who had insisted on packing up the Oval Office himself, raced to meet the deadline. Hillary, as organized and punctual as her husband was chronically tardy (for eight years the administration ran on what was derisively known as "Clinton time"), spent what little time remained walking the halls of the residence. The walls leading to the third-floor solarium, a glassed-in room on the south side of the building, were papered with framed family photographs: a tutu-wearing Chelsea fresh after her performance in *The Nutcracker,* the Clintons sitting at a picnic table, Hillary and Chelsea sharing a hammock. Hillary looked out over the pink geraniums on the terrace, toward the Washington Monument. Next to Chelsea's Beanie Baby collection were several colorfully painted Russian nesting dolls, each fashioned in the image of the Reagans, Bushes, and the Clintons.

The First Lady lingered in Chelsea's rooms, trying to hear in her mind, Hillary would later recall, "the laughter of her friends and the sound of her music. Many of her memories growing up in the White House as the daughter of a President were happy," Hillary added, as if trying to convince herself the good outweighed the bad. "I was sure of that. . . ."

It was around 2 A.M. Saturday when Hillary barreled straight past Democratic National Committee Chairman Terry McAuliffe and into the Oval Office. McAuliffe, a staunch defender of both Clintons who had raised tens of millions of dollars for their various campaigns, was recording the historic final moments of the Clinton administration on a camcorder.

In a matter of hours, the Bushes were to arrive, as was the custom, for coffee with the incumbent before driving off to the Capitol for the inauguration ceremonies. But there, at 2 A.M., stood Bill, crimson-faced and bleary-eyed, surrounded by cardboard boxes and giant rolls of bubble wrap. Despite the fact that his voice was hoarse with exhaustion, Hillary's husband regaled anyone in earshot with the story behind each memento as he tossed them, one by one, into boxes marked WASHINGTON, LIBRARY, and CHAPPAQUA. Whatever didn't go into the boxes was placed on a long table to be picked over by staff and friends. "It feels like a 7-Eleven around here," observed one of the stewards, who watched in amazement as aides and secretaries examined items left out by the President as if they were at a yard sale. In the end, only a pair of presidential pajamas would remain stretched out on the table, too personal an item for anyone to touch, much less take home as a souvenir.

"For God's sake, Bill," Hillary said, interrupting her husband as he continued to wax nostalgic, this time over a photo of him and Hillary taken with Jacqueline Onassis in Martha's Vineyard. "Stop talking and get some sleep!"

The President, shaking his left hand in the air as he winced in pain, ignored his wife. "Man, that hurts," Bill said, grimacing. In this final packing frenzy, the President had managed to slice open his index finger on one of the boxes. Old Arkansas buddy Harry Thomason, who was helping Bill sort through the debris, had tried to close the wound with Super Glue.

"Super Glue?!" Hillary said in amazement, rolling her eyes. It was the just the kind of bizarre and not altogether rational stunt Bill might be expected to pull when he got too tired. Hillary was, in fact, alarmed at just how drained her husband looked—more exhausted than she had ever seen him. Shunning sleep altogether and subsisting on a diet of éclairs, hot dogs, and pizza, Bill had vowed to pack the work of an entire third term into his few remaining days in office. If he could accomplish this, he told Hillary, "it would feel like four more years." Toward that end, in addition to the deal with the Independent Counsel, he made scores of appointments, nominated nine new federal judges, wrote thousands of pages of new federal regulations, and approved the creation of eight new national monuments.

With Hillary's help, Bill also used his last remaining hours in office to compile a list of drug traffickers, fugitives, tax cheats, embezzlers, armed radicals, friends, and relatives who would be granted presidential pardons. That list, along with the hundreds of thousands of dollars' worth of gifts and furnishings they were improperly spiriting out of the White House, would later threaten to bury the Clinton legacy once and for all.

As the administration of William Jefferson Clinton wound down to its final few days, Hillary had watched as her husband wallowed shamelessly in nostalgia, alternately euphoric and melancholy as he passed out coffee mugs emblazoned with the presidential seal, pens, hats, golf clubs, and other souvenirs to anyone who dropped in to say good-bye.

On the Clintons' last night in the White House, Thomason asked if he and the President might do what they did on their first night there in 1993—bowl in the basement alley that had been installed by President Nixon. But that idea was nixed by Hillary. The First Lady, fresh from hosting her press aide's engagement party, wanted to screen the new David Mamet film *State and Main* in the White House theater instead.

After the Clintons watched the movie, Bill went to the kitchen with Thomason and polished off several helpings of apple cobbler. Then the President returned to the Oval Office to resume packing with the help of several junior staffers. He was, they observed, still "running on empty"—so obviously wrung out they feared he might simply collapse.

It was nearly dawn when Bill packed away the last paperweight and the last framed photograph. Hillary had been in bed for hours, but Harry Thomason, who along with his wife, Linda Bloodworth Thomason, had been with the Clintons for their last Thanksgiving and their last Christmas in the White House, stayed up to keep his friend company. In return, Clinton gave Harry his favorite putter.

"Well, better get to bed," Bill sighed to no one in particular. "Last night in the White House . . ." He thought about what he had said for a moment and smiled. Then, trying for Arnold Schwarzenegger but sounding more like Elvis, he quipped, "We'll be *back*."

Six hours later, Hillary stepped into the Grand Foyer with her husband. Ringing the foyer was the permanent household staff, there to bid the Clintons good-bye. From the beginning of their tenure here, the Clintons had rubbed many staff members the wrong way with their lack of punctuality, their oddly imperious manner (at first Hillary instructed staffers not to make eye contact when she passed by), their hair-trigger tempers, their unpredictable hours (Hillary and Bill often rang up the staff at 2 or 3 A.M. to demand something), and their unnerving penchant for shouting Anglo-Saxonisms at the tops of their lungs—at aides, and at each other.

Rankling most of all among old-timers was a lack of decorum that contrasted sharply with the patrician style of the Clintons' predecessors, George H. W. and Barbara Bush. One veteran steward recalled the night eight years earlier when Linda Bloodworth Thomason and another Clinton friend, television actress Markie Post, jumped up and down on the Lincoln bed shouting, "We've made it! We're in the White House now!"

But on this occasion emotions ran high as Hillary moved down the line thanking everyone, from the kitchen staff to the groundskeepers to the maids, for "taking such wonderful care of us each and every day."

The President enveloped each staff member, male and female, in a crushing bear hug—"more of a body-slam, really," observed one breathless recipient of Bill's affection. "I'm really going to miss you," chimed in one of the stewards, "but I hear the next people go to bed at nine." Hillary laughed, then gave White House butler Buddy Carter a lingering embrace that morphed into a waltz. Bill cut in, and the First Couple twirled down the hallway toward the Blue Room. (Still, a number of household staffers would throw their own "good riddance" party to celebrate the Clintons' departure.)

When George and Laura Bush arrived, their predecessors greeted them warmly. "Bush really connects," Bill would later say of this meeting. "It's a mistake to underestimate him." Hillary was not about to make that mistake, now that she would be dealing with Bush from her own position of power on the Hill. At one point, W spotted Chelsea across the crowded room, trying hard not to be noticed as she wiped a tear from her eye. He sidled over to the Clintons' only child and wrapped a reassuring arm around her shoulders. Chelsea beamed and quickly regained her composure. What kept her from breaking down entirely, Chelsea later told a fellow student at Oxford University, was the conviction that her mother would recapture the White House for the Clintons.

As the two first families headed toward the door, they passed a member of the Marine Band seated at a piano in the Grand Foyer. Bill stopped, slid onto the bench next to the musician, and swayed dreamily to the wistful strains of "Our Love Is Here to Stay." Hillary looked at her husband, her features hardening for one fleeting moment, then walked on.

Her collar turned up against the cold, Hillary squinted into the sun and shivered as George W. Bush took the oath of office. Even as a twenty-one-gun salute thundered across the National Mall, workers swarmed over the Oval Office, giving it the "thorough scrubbing" Bush had promised it would get in the wake of the Monica Lewinsky scandal.

No longer President and First Lady, Bill and Hillary headed for Andrews Air Force Base, where Buddy the presidential dog waited for them at the top of the stairs of Presidential Air Mission 28000. A crowd of supporters had gathered inside a hangar to give Hillary and Bill a proper send-off. "I left the White House," Bill told them, "but I'm still here." The hangar erupted in cheers when he turned to Hillary and announced, "You've got a senator over here who will be a voice for you. I'm very proud of her, and I'm very, very proud of Chelsea.

"So we're going on to New York and spend the weekend and then Hillary will show up promptly," Bill said, again gesturing to his wife the senator, "so as not to miss any votes. . . ."

Embodied in this moment was the ritual passing of the torch from one Clinton to another—and the fulfillment of an understanding that had sustained their relationship for three tumultuous decades. Yet behind her familiar toothy smile, Hillary worried that her husband, now left to his own devices, might self-destruct as he had so many times before.

She was not alone. One of Bill's most trusted advisers predicted that his former boss's ego would be crushed and that he'd "definitely go off the deep end." Friends recalled what happened when he lost the Arkansas governor's race in 1980, for example, and was out of office for two years. "Bill basically went crazy sexually," said a close family friend. "We're all terribly afraid it'll happen again."

It had been arranged for the Clintons to leave on a DC-9, but Hillary, ever mindful of appearances, wanted Bill to hold out for one of the two fully outfitted 747s that serve as Air Force One. It was important that the senator's New York constituents be treated to the full presidential spectacle.

When the plane that had been loaned to them was returned to its hangar at Andrews later that day, the maintenance crew was shocked to see that the interior had been stripped bare. The silverware and china bearing the presidential seal, the glassware, condiments, blankets, pillows, candies—even toiletries like toothpaste and mouthwash—were gone. "Thank God," said one dumbfounded crew member, "the seats were bolted down."

The next day, Hillary stayed behind closed doors at their new house in the Westchester County village of Chappaqua, unpacking some of the merchandise they had taken from the White House. Bill, meanwhile, donned a fleece pullover and, with Hillary's brother Hugh following in his SUV, headed out to a local deli. Outside Lange's Little Store, a small group of startled townsfolk who had stopped to gawk began chanting "eight more years." Inside, Bill shook hands with customers while he waited for his order—an egg salad sandwich for himself and a French vanilla/regular coffee for his wife the senator. When Kathleen McAvoy's daughter Siobhan balked at getting Clinton's autograph, McAvoy asked, "Don't you want a President's signature?"

"He's not a President," the little girl responded. Bill, smiling wanly, left with his egg salad sandwich and Hillary's coffee.

On that first Monday following the Clintons' departure from the White House, a station wagon emblazoned with THE MAIDS—AMERICA'S MAID SERVICE pulled up to the Dutch colonial on Chappaqua's Old House Lane and disgorged three women loaded down with carpet sweepers, dust mops, and vacuums. As the maids entered the house, Hillary and Bill, both clad in jeans and parkas, emerged to take the dog for a walk—and satisfy cameramen who had been waiting hours for a photo op. Bill held Buddy's leash with one hand and Hillary's hand with the other, beaming for the cameras and insisting that he was having "a good time unpacking." Hillary made a point of telling reporters that she had gotten up early and conferred with her Washington staff by phone.

As they turned to walk back inside, someone shouted, "Hey, move!" And another, "Get the fuck out!" Bill and Hillary spun around to see that the crowd was cursing at one of their own—a photographer who was blocking their shot. "I thought you were talking to us," cracked Hillary, raising her voice so that it was audible over the din of traffic from nearby Route 117. "How soon they forget." As they ambled up the driveway, Bill threw his arm around Hillary's neck. "You know we *are* going back," he murmured in her ear, wrongly assuming they could not be overheard. Hillary turned and looked up into his eyes.

"We?" she whispered in reply.

Hillary was destined to run the show from the very beginning.

—John Peavoy, longtime confidant

■

There was always the perfectionist, the drive, always the ambition.

—David Rupert, Hillary's first love

When I look at what's available
in the man department,
I'm surprised more women aren't gay.

—Hillary

■

Shit, I can't even get her to use my last name.

—Bill

2

Intent on witnessing his daughter's graduation, Hugh Rodham left his home in the Chicago suburb of Park Ridge the night before, flew to Boston, checked into a motel near the airport, and then boarded the first train to Wellesley. It was important that someone from the family be there; Dorothy Rodham, who had been put on blood thinners and advised by her doctor not to travel, stayed behind in Park Ridge to care for Hillary's younger brothers. Now Hugh watched proudly as Hillary, in her capacity as Wellesley Student Body President, strode purposefully to the microphone.

Chosen to represent the Class of 1969, Hillary was following the day's main commencement speaker, Massachusetts Senator Edward Brooke. Only two years before, Hillary had campaigned for Brooke, a liberal Republican and an African-American, as president of Wellesley's Young Republicans.

But Hillary had changed. Dropping her prepared text, she wasted no time lambasting her predecessor at the podium. "Senator

Brooke," she began, "part of the problem for empathy with professed goals is that empathy doesn't do anything." What her generation wanted now, she said, was action. She ended with a classmate's poem that damned "The Hollow Men of anger and bitterness."

Brooke, obviously singled out as one of the "Hollow Men," was stunned, hurt—and convinced that this was no extemporaneous speech. "As far as I could tell, she was not responding to anything I was saying," he later observed. "She came that day with an agenda, pure and simple."

But Hillary claimed she was reacting viscerally to what Brooke had said. He had mentioned the Vietnam War and growing racial tensions only obliquely; for the most part, Hillary said, his was just another "onward-and-upward" graduation speech. But what really rankled Hillary was her perception that the senator's remarks were somehow pro–Richard Nixon—a call to arms for any self-respecting campus activist in the 1960s.

In response, Hillary offered nothing more than the muddled, sophomoric peace-and-love dogma that was so prevalent on campuses at the time. And, predictably, when it was over, Hillary's mesmerized classmates leaped up to their feet and cheered.

A sizable number of people in the audience were incensed—including short, sullen Hugh Rodham, a dyed-in-the-wool Republican who admitted that at that moment he wanted to "lie on the ground and crawl away." Hillary's father stiffened when he approached her after the ceremony. His reaction hardly surprised her. Even if she had not ambushed the distinguished senator from Massachusetts, Hillary knew her father—unlike the other dads at Wellesley that day—would never throw his arms around his daughter and tell her he was proud of her. Not even when, as a child, she proudly handed him her report card. "It must," he would say, reading down the column of A's, "be a very easy school you go to."

That graduation day at Wellesley, Hillary was embraced by her classmates and even some of her classmates' parents—but not by her own father, whose approval she had always so desperately craved. Hillary would, in fact, always say that it was her self-made dad who spurred her on, simply by holding out the promise of his affection as a reward for high achievement. After four years at Wellesley carving out an identity of her own, however, it was dawning on Hillary that her father's love might never be forthcoming. In a description fraught with Freudian overtones, she would later describe her father as a "self-sufficient, tough-minded small businessman."

No matter. Once her father departed for home, she ran to Wellesley's Lake Waban, doffed her graduation gown to reveal a bathing suit underneath, and—in violation of the college's strict rule against swimming in the lake—dived in. When she emerged, her clothes were gone. Wellesley's president, Ruth Adams, had spotted Hillary swimming and, seething over the sneak attack on Senator Brooke, ordered security to confiscate them.

Adams was not alone. Hugh Rodham fumed about his daughter's impertinent remarks all the way back to Park Ridge. Sending his only daughter to Wellesley in hopes that she would receive a traditional finishing school education was, Rodham conceded only half-jokingly, "a great miscalculation!"

If, by withholding his love, Hugh Rodham lit a fire under Hillary, it was Dorothy Howell Rodham who stoked that fire with affection and encouragement—and told Hillary from the age of seven that she should aim for a seat on the U.S. Supreme Court. Dorothy's own childhood had been anything but idyllic. Hillary's Welsh-English grandfather was seventeen and an apprentice firefighter in the slums of South Chicago when Dorothy was born in 1919. Dorothy's French-Scottish mother, Della Murray, was just fifteen—and illiterate.

When she was eight years old and her sister only three, Dorothy Howell's parents divorced. The two girls, terrified and alone, were put on a train bound for California—a harrowing three-day journey that Dorothy would never forget. Hillary would later say that every time her mother mentioned the cross-country train trip, she was "furious that any child could be treated like that." Things only got worse when they settled in with their grandparents, British immigrants who were both physically and emotionally abusive to the little girls placed in their care.

These Dickensian visions of small children being cast out to fend for themselves served as an object lesson for Hillary, who was taught from the cradle to believe that divorce was disaster. "Children without fathers," she would later write, "or whose parents float in and out of their lives after divorce, are precarious little boats in the most turbulent seas."

Dorothy was fourteen when someone finally tossed her a lifeline, offering her a job as a live-in babysitter for a local family. Away from the poisonous atmosphere of her grandparents' home, Dorothy flourished. At Alhambra High School, she joined several student organizations—the Spanish Club, the Scholastic Society, the Girls Athletic League—and excelled both academically and athletically. Graduating in 1937, she returned to Chicago and took a job as a secretary—"It's what you did if you were a woman back then," she later explained—at the Columbia Lace Company. Two years later, she met and began dating a young salesman named Hugh Rodham.

Like Dorothy's, Hugh's childhood had been marred by ignorance and poverty. Hillary's paternal grandparents were Welsh immigrants who, in an era before child labor laws, settled in Scranton, Pennsylvania, and went to work rather than attend school. Hugh landed a football scholarship to Penn State, and after graduating with a degree in physical education went to work unloading crate boxes at a warehouse. Later, Hugh struck out for Chicago—and a salesman's job at the company where Dorothy Howell worked, Columbia Lace.

Dorothy and Hugh were married after a five-year courtship, in 1942. Almost immediately, Rodham enlisted in the navy and discovered a unique opportunity to put his degree in physical education to good use. Rodham was assigned to whip raw recruits into shape using the Gene Tunny program, a regimen devised by the retired world heavyweight boxing champion.

When he returned to Chicago after his discharge, Hugh declined an offer to return to his old job. Instead, recognizing that a postwar housing boom would mean a surge in demand for home furnishings, he launched his own custom drapery business. He and Dorothy were ensconced in a tiny one-bedroom apartment in Chicago's Lincoln Park district when, on October 26, 1947, Dorothy gave birth at nearby Edgewater Hospital to eight-and-a-half-pound Hillary Diane. Dorothy chose what she had always believed to be a man's name, Hillary, because to her it sounded "exotic."

Even as a toddler, eager-to-please Hillary impressed her mother as being "very mature, very grown up." When Hugh Jr. arrived three years later, the family relocated to suburban Park Ridge, an upscale, all-white Republican stronghold thirty-five miles northwest of Chicago.

A tidy, two-story brick house encircled by shade trees at 236 Wisner Street—the corner of Wisner and Elm—would be the Rodham family home for the next thirty-seven years. For Dorothy, who had given up her own dream of attending college to fulfill the classic 1950s role of happy homemaker, the dignified-looking house with the arching windows and flagstone facade was also a prison.

Hugh also played his role—that of the gruff, career-obsessed, tobacco-stained, crabgrass-battling dad—to perfection. But he went a step further. Although he indulged himself with a brand-new Cadillac every year, he was unsparing with his wife and children. Rodham paid Hillary and her two brothers (Tony arrived when she was eight) one penny for every weed they yanked out of the yard. Hillary woke up shivering every morning because her father turned off the heat at night. He swore "a blue streak," as one neighbor put it, if things weren't done *just his way.*

Beneath the *Leave It to Beaver* veneer was the simmering domestic discontent familiar to many children of the 1950s. In the evenings, Hillary hid in her room while her parents hurled invective at each other over cocktails. Whatever the degree of her frustration, Dorothy, in keeping with the mores of the time, would never dream of airing her marital grievances in public. Yet she was also determined that "no daughter of mine was going to have to go through the agony of being afraid to say what she had on her mind. Just because she was a girl didn't mean she should be limited."

No sooner had the Rodhams arrived in Park Ridge than four-year-old Hillary was confronted with someone hell-bent on "limiting" her options. A local girl named Suzy was the scourge of the neighborhood, routinely pummeling both boys and girls with unabashed glee. When Hillary came sobbing to her mother that she was afraid of Suzy, Dorothy Rodham offered no words of comfort. "There's no room in this house for cowards," she told her daughter in what would be a turning point in Hillary's childhood. "The next time she hits you, I want you to hit her back."

Hillary marched back to Suzy's house and, with an audience of boys on hand to witness her revenge, slugged the unsuspecting Suzy square in the face. Hillary, beaming with pride, dashed straight back home to tell her mother. "I can play with the boys now!" she proclaimed.

"When she was old enough to play outdoors by herself," Dorothy later recalled, "she could beat up on the neighbors' children, but only if she had to. When she did, she'd go out, arms flailing, eyes closed—and whap! She'd get the better of them." In a neighborhood where boys outnumbered girls two to one, Hillary had few female playmates. Still, whether the game was hide-and-seek, chase-and-run, or cops-and-robbers, Hillary invariably ran the show. "Boys responded well to Hillary," Dorothy recalled. "She just took charge, and they let her."

Yet the one person whose approval Hillary craved the most "was never satisfied," Dorothy later conceded. The Rodham children's Norman Rockwell childhood of skinned knees, bike races, lost skate keys, and kiddie matinees was tainted by Dad's forbidding presence. He made Hillary memorize stock quotes as well as baseball statistics, and when she couldn't hit a curveball to his satisfaction, Rodham took her to Hinckley Park near their home and pitched balls at her for hours at a time until she could. On those rare occasions when she misbehaved, it was Dad the stern disciplinarian who threw her over his knee and spanked her.

To those in the neighborhood, Hugh was a dour, unsociable-to-the-point-of-rude character who never answered the front door or even bothered to acknowledge the presence of visitors to his home. Hillary would always remember the day her long-suffering mother took out a carpenter's level and used it to give her some pointers on how to remain centered. Dorothy told her daughter to imagine that the carpenter's level was inside her, and then she tipped it so that the bubble went to one end, then the other. "You try," she said, "to keep the bubble in the center."

The advice paid off for Hillary, who was a model student at Eugene Field Elementary School—she spent her after-class hours covering her Girl Scout sash with the most merit badges of any girl in her troop—and then at Ralph Waldo Emerson Junior High. Hillary's peers weren't sure what to make of her. According to classmate Betsy Johnson, the "other girls would say, 'Oh, she's *so* conceited.' And I think it wasn't until we were in high school that we realized what they took for conceit in Hillary probably was this sense of self-confidence that she'd always had. Always."

Park Ridge nurtured overachievers like Hillary. "It was a very conservative town," recalled one neighbor whose father, like so many residents of Park Ridge, was a member of the right-wing John Birch Society. "Kids tried to please their parents back then, and nobody tried harder than Hillary Rodham."

Not that she had much choice. Her grumbling, grousing father brooked no disagreement, especially when it came to politics.

Mom sat silently while her husband railed against FDR, Harry Truman, Adlai Stevenson, and the Kennedys, and when she accompanied him to the voting booth, he assumed she was following his lead and voting for Richard Nixon against John F. Kennedy. She wasn't. Mrs. Rodham may have urged her daughter to stand up for herself, but Dorothy's sole act of defiance was to become a secret Democrat.

Hillary kept trying to earn her father's approval at Maine Township public high school, performing in school musicals and plays, winning class offices, working on the school paper, joining clubs (the pep club, the debating team, the brotherhood society), and racking up scholastic awards. She was a National Merit Scholar—one of only eleven at her school of 1,400 students. "She was ambitious as hell," said one classmate, who said Hillary talked constantly about what would "look good on my résumé." Another student, Arthur Curtis, agreed: "Hillary was very competitive at everything." Curtis was taken aback when Hillary told him, "I'm smarter than you." He was not alone. Where other children were told that it was not nice to brag, Hillary routinely informed her classmates that she was the smartest student at Main Township High. "Hillary was taught to fight," Curtis said, "but she was never taught *manners*."

None of it seemed to matter much to Hugh Rodham, who only grudgingly agreed to buy her a dress for the junior prom because Dorothy was going to be a chaperone and she didn't want to be embarrassed. Dorothy was, in fact, concerned about Hillary's appearance—Hillary irked her mother by refusing to wear makeup—and her apparent lack of interest in boys. When the school newspaper predicted that the humorless, compulsive overachiever would wind up in a convent as "Sister Frigidaire," Hillary paid little mind. "She thought it was all superficial and silly," Dorothy said. "She didn't have time for it."

Indeed, as a teenager Hillary seemed hell-bent on filling every spare moment with fresh ideas and eye-opening experiences.

Through her church, First United Methodist, she volunteered to babysit the children of migrant farmworkers brought in each year to work the fields not far from Park Ridge. The Reverend Don Jones, a social reformer whose own guiding philosophy was anathema to Hugh Rodham, took Hillary and other members of his University of Life youth group to visit black and Hispanic inner-city churches.

In April of 1962, Jones told Hillary and her church that they were going to Orchestra Hall to hear Martin Luther King deliver a speech called "Sleeping Through the Revolution." Many of the other students were forbidden to go; in Republican Park Ridge, King was viewed as a rabble-rouser. But with a little gentle prodding from Dorothy, Hugh grudgingly signed the permission slip. After the lecture, Jones took Hillary and the others backstage to meet Reverend King—a moment that would be indelibly etched in Hillary's memory.

So, too, was the day when a teacher burst into Hillary's high school geometry class to announce that President Kennedy had been gunned down in Dallas. "Probably some John Bircher," her geometry teacher muttered before instructing everyone to file into the auditorium and wait to be sent home. When Hillary arrived, Dorothy, sitting spellbound before the family television set, admitted for the first time that she had voted for JFK.

These eye-opening events notwithstanding, Hillary would admit that she continued to parrot Hugh Rodham's beliefs. Hillary devoured Barry Goldwater's *Conscience of a Conservative* and wrote her term paper on the American conservative movement. Already an active member of the Young Republican Club, she went a step further and signed on to campaign for the Republican presidential candidate as a "Goldwater Girl"—right down, she would later admit, to the cowgirl getup and the hat bearing the slogan AuH_2O.

Around this time, Hillary decided to make her first run for "The Presidency," as she solemnly referred to it—of her senior class. She had already served as vice president of her junior class and was eager to take on the top job. But the two boys she was running against made it clear they did not take the idea of a female candidate seriously. "One of the boys told me," she later said, "'You're really stupid if you think a girl can be elected president.'" Hillary was soundly defeated on the first ballot—"which didn't surprise me but still hurt," she would recall forty years later.

It was not the first time she had been thwarted by sexism; when she wrote NASA saying she wanted to be an astronaut, she received a letter from the agency coolly informing her that there were no plans to train women for careers in space. But the race for senior class president did expose Hillary for the first time to what, even then, she referred to as "dirty politics." She called up Reverend Jones to complain bitterly of her opponents' "mud-slinging"—and vowed never to take the high road again if it meant losing an election. "It was a bitter pill for her," he said. "She was deeply hurt—and angry. Hillary hated to lose."

She couldn't be senior class president, but Hillary was determined to be a highly visible presence on campus. In addition to campaigning as a Goldwater Girl, Hillary proposed holding a mock political convention in the school gym. The teacher who oversaw the "convention" knew that Hillary was campaigning for Goldwater, just as he knew Hillary's friend Ellen Press was a supporter of incumbent Lyndon Johnson. To make things more interesting, Press was given the task of representing Goldwater while Hillary played LBJ. "I resented every minute of it," she recalled.

Hillary was voted Most Likely to Succeed when she graduated from Maine Township High in 1965. That fall Dorothy and Hugh made the grueling eight-hundred-mile drive to Wellesley for the first time—somehow managing to get lost in Boston and ending up in Harvard Square. Harvard was teeming with shaggy-maned radicals and scruffy potheads—or at least that's the way it looked to Hugh Rodham. He threatened to turn back. But when the

Rodhams finally did find their way to Wellesley, Hillary's father was relieved at what he saw. There were no bearded hippies; in fact, with the exception of the stray tweed-jacketed faculty member, there were no men at all.

A collection of brick-and-stone neo-Gothic buildings sprinkled across five hundred wooded acres, Wellesley was regarded by many as the country's most beautiful college campus. The sylvan setting was an important part of Wellesley's genteel image. Since its founding in 1875, Wellesley (like the other Seven Sisters—Barnard, Bryn Mawr, Mount Holyoke, Radcliffe, Smith, and Vassar) catered to the pampered daughters of America's privileged elite.

The college's unsurpassed academic reputation, the manicured grounds, the status, the contacts, the sense of tradition—the most famous of these involved rolling hoops into Lake Waban to see which Wellesley grad would be the first to marry—all were factors in Hillary's decision to attend Wellesley. One reason eclipsed all the others: Hillary had chosen Wellesley precisely because it was so far away from her autocratic father.

Nevertheless, as she watched her parents drive away, Hillary felt "lonely, overwhelmed, and out of place." Most of her classmates had gone to boarding schools, vacationed every year in places like Palm Beach and the Côte d'Azur, and spoke several languages. They also seemed to have the edge academically. Hillary excelled in what would become her major, political science, but faltered in geology, math, and French. After a month, she placed a collect call to Park Ridge and told her parents, "I'm not smart enough to be here."

Hugh Rodham had never praised his daughter for getting into the exclusive college and would just as soon have paid for Hillary to attend a cheaper school somewhere in the Midwest. He told her to come home. Dorothy, however, insisted she stay. "Don't be a quitter," she said. "We're not quitters."

Hillary remained at Wellesley, focusing on her own ambitious scheme to restructure Illinois's state Republican organization so that by the time she graduated she could make a serious run for office. Hillary also threw herself into her studies and class activities, taking over the Young Republicans and winning a seat in the student senate.

Hillary had a room to herself in the Stone-Davis dormitory, a neo-Gothic structure perched on a hill with breathtaking views of Lake Waban. She would remain at Stone-Davis for all four of her years at Wellesley, dining each day with her friends in a flower-filled, glass-walled gazebo. Already adept at networking, Hillary quickly determined who was important to know on campus—and who might be of use to her in the future. Among the friends she would make during this period: Teddy Roosevelt's great-granddaughter Susan Roosevelt, who would go on to marry future Massachusetts Governor William Weld, and Eleanor "Eldie" Acheson, granddaughter of Truman's secretary of state, Dean Acheson.

With no men on campus to impress during the week, the women of Wellesley paid little attention to their appearance—and Hillary was no exception. Her dyed-blond hair had dulled to a lifeless brown, and she pulled it back in a ponytail or schoolmarmish bun. Hillary still eschewed cosmetics, and her bottle-bottom glasses were thicker than ever. Her wardrobe was pure sixties—tie-dyed shirts, frayed jeans, beads, and sandals.

Hillary and her classmates did make more of an effort to look presentable on weekends, when they took the train to Cambridge to go out on dates with Harvard boys. Since Wellesley had a 1 A.M. curfew on weekends, Hillary would later remember that Route 9 between Cambridge and Wellesley was "like a Grand Prix racetrack . . . as our dates raced madly back to campus so we wouldn't get in trouble."

Hillary soon began dating Geoffrey Shields, a prelaw student at Harvard who hailed from another upscale Chicago suburb, Lake Forest. Hillary went with Shields on hiking trips and to the occasional football game, but Hillary never seemed more fully engaged than when they were seated with friends on the floor of Shields's Harvard dorm debating the issues of the day: poverty, civil rights, the Vietnam War. "That," said Shields, "is when Hillary really came alive."

In the beginning, Hillary defended the presence of U.S. troops in Vietnam. But as a sophomore, she underwent a change of heart. Realizing that her beliefs "were no longer in sync with the Republican party," she resigned as president of the Young Republicans. Not that she could be remotely described as a radical. Determined to bring about change by working within the system, Hillary pressed for greater minority enrollment—Wellesley counted only ten blacks among its students at the time—as well as an end to curfews.

As confident as she was as an advocate, Hillary harbored doubts about the course her own life should take. Long before her future husband proclaimed, "That depends on what the meaning of the word '*is*' is," Hillary asked, "I wonder who is me?" Writing to John Peavoy, a high school classmate now attending Princeton, Hillary mulled over which identity was right for her: that of "educational and social reformer, alienated academic, involved pseudo-hippie, or compassionate misanthrope."

Peavoy viewed Hillary's soul-searching as "typical of the time, and also typical of the age. With the Vietnam War and the civil rights movement in full swing, there was no way you could *not* be involved. We were still adolescents, really. So it was not so much 'These are tumultuous times' as it was 'How does all of the this affect *me*?' "

Ultimately, Hillary did not choose reformer, academic, pseudo-hippie, or misanthrope. Instead she settled on a fifth option: politician. "From the very beginning," Peavoy recalled, "there was never any doubt that she was going to be the leader, at the head of something big."

Although she wrote to Peavoy saying that she did not regard

herself as one of the "faceless masses," Hillary felt it was her duty as a committed Methodist to lead her fellow citizens down the right path. With the pain of her high school election defeat still fresh, Hillary nonetheless decided to run for Wellesley student government president. To her amazement, she won.

Hillary set out immediately to push her own agenda, lobbying hard for an end to Wellesley's mandatory curriculum—a loosening of academic requirements that, years later as the parent of a college student, she would come to regret. She also campaigned successfully to end college rules barring men from setting foot in Wellesley dorm rooms.

At twenty, Hillary was already being criticized for using her office to reward cronies with assignments to key school committees—a charge that would be leveled against her and her future husband repeatedly over the coming years. "The habit of appointing friends and members of the in-group should be halted immediately," demanded the *Wellesley News,* "in order that knowing people in power does not become a prerequisite to office holding."

Politically, Hillary's metamorphosis from Goldwater Girl to student activist was continuing apace. By the time she was a junior, Hillary was waving placards at antiwar rallies and chanting "Hey, hey, LBJ, how many kids have you killed today?" In March 1968 she drove up to Manchester, New Hampshire, to campaign for Eugene McCarthy, the antiwar Minnesota senator who was challenging LBJ in the Democratic presidential primaries.

Hillary was buoyed by McCarthy's strong showing in New Hampshire, and by the subsequent entry of Robert F. Kennedy into the race. But perhaps the single most pivotal event in Hillary's political transformation was the assassination of Martin Luther King on April 4, 1968. Upon hearing the news, Hillary became hysterical. Once she regained her composure, she called her black friends at Wellesley to commiserate, then organized a small group to march with demonstrators at Post Office Square in Boston.

Later Hillary, frustrated by Wellesley's business-as-usual atmosphere, organized a two-day campus strike.

Bobby Kennedy's assassination just two months later intensified Hillary's growing sense of despair and bewilderment. Yet she went ahead with plans that summer to intern in Washington, even though it meant reporting to the House Republican Conference. The group was then headed by Minority Leader Gerald Ford, and Hillary found herself working closely with New York Congressman Charles Goodell and Melvin Laird of Wisconsin. Hillary grew especially fond of Mel Laird, who would later serve as President Nixon's defense secretary. Although they disagreed about the conduct of the Vietnam War, Laird took Hillary and the other interns seriously and actively sought out their opinions. (In stark contrast to a President whose exploitation of White House interns would lead to impeachment, she would later recall that Laird and the other congressmen she encountered treated women as equals. "I have pretty good antennae," Hillary said the day before the Lewinsky scandal erupted, "for people who are chauvinist or sexist or patronizing toward women.")

At the end of her internship, Hillary was asked by Goodell to go to the Republican convention in Miami and work on behalf of New York Governor Nelson Rockefeller's eleventh-hour campaign to snatch the GOP nomination from Richard Nixon. Hillary would later say this first look into "big-time politics" was "unreal and unsettling"—for reasons that went beyond the power struggle between Nixon and Rockefeller.

For starters, it marked the first time that Hillary, whose tight-fisted father would never spring for something so extravagant, actually stayed at a hotel—Miami's fabled Fontainebleau—and ordered room service. During the convention, Hillary also got to meet Nixon supporters Frank Sinatra and John Wayne, both of whom "feigned interest" in meeting her.

Hillary's contact with another intern, David Rupert, was less fleeting. The intense, argumentative, darkly handsome Georgetown University government major was working for Congressman Goodell, and he and the girl from Park Ridge hit it off instantly. Within a matter of weeks they were lovers. Hillary could be surprisingly spontaneous when it came to sex, Rupert recalled, but she was never one to risk an unwanted pregnancy in the heat of passion. Hillary always insisted on using birth control whenever they slept together.

"It was an intense love affair," Hillary's friend Nancy Pietrefesa said. "Hillary was always attracted to arrogant, sneering, hard-to-please men, like her father." Hillary's highly charged relationship with Rupert, which she hid from some of the most important people in her life, would last for three years. It included parties at which Hillary presumably did her fair share of inhaling.

Nixon's nomination by the Miami convention hardly surprised Hillary; it had been all but a foregone conclusion. The Democratic convention in Chicago was another matter. When Hillary saw news reports of protesters flooding the streets of the city, she and a friend, Betsy Johnson, were determined to be part of the action. Telling their parents they were off to the movies, the two young women jumped in the Johnson family station wagon and headed for Grant Park, the center of the protests.

A rock hurled by a protester screaming profanities narrowly missed Hillary's head. But like other student activists, she preferred to be "shocked by the police brutality we saw."

Back at Wellesley, Hillary decided to write her thesis on the work of Saul Alinsky, the leftist firebrand whose 1947 book, *Reveille for Radicals,* was regarded by many as the bible of the protest movement. Colorful, outspoken, and often outrageous, Alinsky believed the only way to effect change was by confronting those in power—with protest marches, strikes, and sit-ins.

Hillary was an ardent admirer of both Alinsky and Marxist theoretician Carl Oglesby, who denounced America's "ruling class" and had nothing but praise for Ho Chi Minh, Castro, and Mao.

While she never took issue with their goals, Hillary did not agree with their assertion that change could only be initiated from the outside. For her trenchant analysis of Alinsky and Chicago's Community Action Program, part of the larger War on Poverty, Hillary received an A-plus. Hillary's political science professor, Alan Schecter, deemed all her work not merely insightful but "brilliant."

Alinsky was so impressed with Hillary that he offered her a chance to work with him after graduation, but she turned him down. Although he told her she was throwing her life away—and the chance to make a real difference in the lives of the poor and disenfranchised—Hillary applied to several of the country's top law schools. "The only way to make a real difference," Hillary countered, "is to acquire power."

Accepted at both Yale and Harvard law schools, Hillary was having difficulty choosing—until an imperious Harvard professor stated flatly, "We don't need any more women at Harvard." She had already been leaning toward Yale, but that encounter, Hillary said, "removed any doubts about my choice."

She may have questioned Saul Alinsky's overall strategy, but she embraced many of his tactics. The agitator emeritus believed in a win-at-all-cost approach in the battle for power, and that that required zeroing in with laserlike intensity on one's enemies. Advised Alinsky: "Pick the target, freeze it, personalize it, and polarize it."

Senator Edward Brooke was Hillary's first major target, and the speech in which she castigated him caught the eye of the national media. No less an authority than *Life* magazine proclaimed her one of the eloquent new voices of a restless generation.

Looking for a little adventure that summer following graduation, Hillary escaped to Alaska and a job gutting salmon in a makeshift processing plant that had been set up on a Valdez pier. She was soon promoted to the assembly packing line, where after several days she began noticing that the fish looked "weird. They're black," she told her less-than-amused foreman. "Maybe they aren't fit to be eaten." She was fired on the spot, and told to return the next day to pick up her check. When she did, the entire operation had been dismantled.

In the fall of 1969, Hillary went from an all-female environment to a law school where, out of 235 students, she was one of only twenty-seven women. Hillary made a conscious choice not to change her appearance for the purpose of attracting the opposite sex; she was determined to fit in as the male students' intellectual equal. That meant she still shied away from makeup, paid little attention to the state of her hair, and refrained from shaving her legs. Her wardrobe now consisted of several pairs of denim bell-bottoms, paisley-printed peasant blouses, sandals, black silk pajama bottoms, and of course her trademark thick glasses—a dozen pairs, ranging from wire-rimmed to aviator to oversized red plastic frames.

"She didn't want to be thought of as pretty," a fellow student recalled. "She wanted to be thought of as smart. And she didn't particularly want the distraction of a boyfriend—not when there were so many important things going on in the world. She wanted to be *involved,* but not necessarily with a man."

No sooner did she sign up for classes than Hillary introduced herself to the leaders of the protest movement at Yale. With unrest growing on the nation's campuses, she was intent on playing a key role. Hillary got her first break in April of 1970, when the murder trial of Black Panther founder Bobby Seale was about to start in New Haven. Seale and seven other Panthers had been charged with ruthlessly killing one of their own, Alex Rackley, but not before torturing, beating, and scalding him. Afterward, said police, they mutilated the body. The Panthers had suspected Rackley of being a police informant, and believed he had tipped off the authorities to a New York bombing conspiracy.

New Haven braced for rioting as thousands of angry Panther supporters flooded into New Haven. Another Panther leader, convicted cop killer Hughie Newton, was freed from San Quentin in

California on $50,000 bail and showed up to support Seale, calling for full-scale revolution against "Fascist Amerika." Jane Fonda also arrived in town to whip up the crowd, raising her hand in a Black Power salute.

Like many members of what was now called the New Left, Hillary admired both Newton and Seale. (Seale had already gained fame as one of the Chicago Eight, the group tried for leading the disruptions at the Democratic National Convention.) She suspected that Seale had been set up by the FBI and, doubting whether he could ever receive a fair trial, prepared to take part in a huge May Day rally in support of the Panthers. Hillary remained sympathetic to the Panthers, even when their supporters were suspected of setting fire to Yale's International Law Library. While she joined a bucket brigade of faculty and students to douse the flames, Hillary was thinking of ways to aide Seale and his cohorts.

In their trademark black leather uniforms, storm-trooper boots, dark glasses, and black berets, the Panthers cultivated a swaggering, menacing image that, according to one former member, appealed to the "strong masochistic streak" in the New Left. The party had been formed out of an Oakland, California, street gang in 1967 by Seale and Huey Newton. Panthers, many of whom had done prison time for serious crimes, brandished weapons and patrolled the streets in armed cadres, focusing on cases of police brutality that they said proliferated in the ghetto.

Although it was Stokely Carmichael who coined the slogan "Black Power" and booted whites out of the Student Nonviolent Coordinating Committee (SNCC), the Panthers were the first to blatantly reject the notion of peaceful protest. They also differed from other reform-minded groups such as the Congress of Racial Equality (CORE), King's Southern Christian Leadership Conference, and the Urban League in that theirs was an openly Marxist organization with a forthright revolutionary agenda.

The Panthers had remained a local Bay Area phenomenon until

October 1967, when a single bloody incident thrust them center stage. Newton, the party's self-proclaimed "minister of defense," was leaving a party celebrating the end of his probation for a knifing incident when he was stopped by Oakland policeman John Frey. There was a struggle, and within minutes Frey was dead—the victim of five gunshot wounds, including two in the back at close range. A backup officer was wounded, as was Newton.

With the help of such fellow Panther ideologues as Seale and convicted rapist turned *Soul on Ice* author Eldridge Cleaver, the Newton trial became a cause célèbre. At Yale and other campuses, posters went up on dormitory walls showing Newton sitting on a rattan throne, a rifle in one hand, a spear in the other. Hillary was among the thousands of students who proudly wore buttons that demanded that California authorities FREE HUEY.

Newton's defense team would argue that there was a distinct possibility that the backup officer, not their client, had actually shot his partner in the scuffle. In what would soon become a standard tactic, they contended it was not Newton who was on trial but the *system*.

Newton was found guilty of manslaughter, but the conviction was overturned on a technicality. The new darling of the Left, Newton went to live in a glass-walled penthouse overlooking Oakland's Lake Merritt. There he and Seale held court, guzzling vodka and expounding endlessly on the coming revolution before a rapt audience of students, journalists—and several of Hollywood's leading directors, screenwriters, and actors.

Hillary sympathized with the Panthers, and saw them as a legitimate political force to be reckoned with. They were, in fact, fast becoming a criminal menace. In addition to dealing drugs and taking protection money, the Panthers would be involved in numerous shoot-outs with police across the country.

As for the handsome, charismatic Newton, he turned out to be a chronic alcoholic and abuser of drugs. In 1974, he would flee to

Cuba after being accused of fatally shooting a seventeen-year-old prostitute in the face because she failed to recognize him and of pistol-whipping a tailor for affectionately calling him "baby." Three years later, Newton returned to face the charges, which were dropped after both trials ended in hung juries. Later, he served time for a parole violation and for misappropriating funds raised by the Panthers for one of their Oakland community projects.

In 1984, Newton received a Ph.D. from the University of California at Santa Cruz, but only after allegedly threatening to kill his professor if he didn't receive passing marks. And there would be continuing skirmishes with the law until August 1989, when Newton was shot to death after being locked out of an Oakland crack house.

In 1970, however, the Black Panthers were lionized by Hillary and her like-minded friends. It was a cocktail party held in the Panthers' honor at the Manhattan apartment of Leonard Bernstein that formed the basis of Tom Wolfe's bestselling book *Radical Chic.*

Chic was not the word to describe Hillary Rodham—far from it. She was practical, focused, and, though she would scrupulously avoid mentioning it in later years, committed to doing whatever she could to aid Bobby Seale and his fellow Panther defendants in their murder trial.

Before she could decide on a course of action, however, four students were gunned down during antiwar demonstrations at Kent State. On hearing the news, Hillary rushed out of the law school in tears. A few days later, she traveled to Washington to speak at a banquet marking the fiftieth anniversary of the League of Women Voters. Wearing a black armband, she scarcely kept her emotions in check as she railed against the war, Richard Nixon, and capitalist America. "How much longer can we let corporations run us?" she demanded. "Isn't it about time that they, as all the rest of our institutions, are held accountable to the people?"

When she returned to Yale, Hillary signed up for a project begun by one of her professors, outspoken leftist Thomas (cheerfully referred to by students and faculty alike as "Tommy the Commie") Emerson. Hillary was assigned the job of making certain that there would always be a law school student on hand at the Panther trial to monitor the proceedings and point out any civil rights abuses on the part of prosecutors to the American Civil Liberties Union.

Hillary and her fellow students did not confine themselves to compiling information for the ACLU, however. Emerson introduced Hillary to famed radical lawyer Charles Garry, a member of the Panther defense team, and soon her band of student watchdogs were feeding whatever information they gleaned directly to the Panther attorneys. In later years Hillary would take pains never to mention her friendship with Garry, whose clients included People's Temple founder Jim Jones and Angela Davis as well as Newton and Seale.

Incredibly, Hillary dismissed out of hand the evidence against Garry's clients, which included signed confessions from two of the defendants and a chilling audiotape of the victim's "trial" by his fellow party members before he was summarily executed.

Nor did it seem to matter to Hillary that the Panthers were waging a campaign of intimidation directed at Yale and the surrounding community. "If Bobby dies," Bobby Seale's supporters chanted as they marched through the campus, "Yale fries!" At one point, Panther David Hilliard showed up at a campus rally to proclaim "there ain't nothing wrong with taking the life of a motherfucking pig."

In her memoirs, Hillary would praise Yale President Kingman Brewster for his leadership during this period. That "leadership" consisted of shutting down classes and throwing open the doors of the university to demonstrators. Basically calling for the Panthers' acquittal regardless of the evidence, the esteemed Brewster expressed doubts that radicals could receive a fair trial anywhere in the United States.

Hillary also admired William Sloane Coffin, Yale's left-leaning chaplain and a luminary of the antiwar movement. Just six years earlier, an undergraduate named George W. Bush was devastated when Coffin had unkind words to say about his father, who had just run for the Senate in Texas and lost. Now Coffin declared that the Panthers' "white oppressors" should back off—that it was "morally wrong for this trial to go forward."

By the end of her first year at Yale, Hillary was already known around campus as a major voice in the student antiwar movement. And while former friends and classmates would later claim they did not regard her as a true radical, her outspoken and highly effective support of those causes indicated otherwise. "You've got to remember that when people say they don't remember her as a radical," says one former antiwar activist, "it's because *they* were probably Maoists or worse. Consider the source. Hillary was a radical, all right, but a very businesslike radical."

She had already begun making important contacts in the nation's capital. While in Washington to deliver her fiery speech to the staid League of Women Voters, Hillary had met perhaps her most important mentor, noted civil rights lawyer and children's rights pioneer Marian Wright Edelman.

A 1963 graduate of Yale Law School, Edelman was the first black woman admitted to the bar in Mississippi and headed up the NAACP legal defense fund in that state. Over the next several years, she risked her life organizing voter registration drives and demonstrations to protest segregation. By the time Hillary got to know her, Edelman, whose husband Peter was once an aide to Bobby Kennedy, had used her considerable leverage to establish the Washington Research Project in D.C. The project would soon evolve into the Children's Defense Fund.

Hillary, motivated in part by stories of her mother's horrendous childhood, signed on to work with Senator Walter Mondale's subcommittee studying migrant labor. Hearkening back to her own experience babysitting the children of farmworkers in Illinois,

Hillary interviewed scores of laborers about conditions in migrant labor camps.

Even before the hearings started, Hillary was seething. Minute Maid, which had just been acquired by the Coca-Cola Company, was one of the companies targeted by the investigation. On Capitol Hill, Hillary waited patiently for the arrival of Coke president J. Paul Austin, who was scheduled to testify. As soon as she spotted him, Hillary pointed a finger at the hapless executive. "We're going to nail your ass," Hillary said point-blank. "Nail your ass!"

Later, Hillary would work with the staff of Yale–New Haven Hospital drafting legal guidelines for the medical treatment of battered children, and write papers on the legal rights of minors for the nonprofit Carnegie Council on Children. Years later, Hillary would be wrongly accused of advocating changes in the law that would allow children to sue their parents if they didn't want to take out their garbage—a misconception she would attempt to correct in her folksy, upbeat, and cautiously moderate book *It Takes a Village.*

But Hillary's earlier writings may more accurately reflect her true beliefs because they are not intended to mollify a wider audience. They are crafted in the strident prose of the committed social engineer, and make an explicit argument for the state to play a more active role in child rearing. Referring to children as "political beings," Hillary challenged the autonomy of the family. "The pretense that children's issues are somehow above or beyond politics endures," she complained, "and is reinforced by the belief that families are private, nonpolitical units whose interests subsume those of children."

Hillary did, in fact, argue that children be given fundamentally the same rights in court as their parents—including, if the need arose, the right to sue them. "Ascribing rights to children," wrote the woman who would one day bar her own teenage daughter from getting a tattoo, "will force from the judiciary and the legislature institutional support for the child's point of view."

During her stint in New Haven's Legal Services office, Hillary was taken under the wing of a young legal aid lawyer named Penn Rhodeen. Hillary helped Rhodeen represent one black foster mother who wanted to adopt a two-year-old girl she had raised since birth. Connecticut had a strict policy, however, that barred adoption by foster parents. Despite their best efforts, Rhodeen and Hillary lost the case, and the little girl was taken from the only mother she had ever known.

Hillary was "passionate" on the subject of children's rights, Rhodeen said. But no one was under the impression that she would be willing to toil in obscurity as a legal aid lawyer. Hillary already had her eye on several top law firms in Washington and New York.

In the meantime Hillary, by now a bona fide star on campus, basked in the adulation of her fellow students. "We were simply," one said, "awed by her. She was so forceful, so self-assured that when she just took charge you accepted that it was the natural order of things."

Hillary would meet her match in that turbulent fall of 1970 when she spotted an orange-bearded "Viking" holding forth in the student lounge. The tall, scruffy-looking character was draped over a sofa and boasting loudly that, for starters, Arkansas grew the world's biggest watermelons.

Bill Clinton, Hillary soon learned, was a Rhodes scholar who had just returned from two years at Oxford. Clinton, in turn, asked their mutual friend Robert Reich what he knew about this serious girl with the Mr. Magoo glasses. For the next several weeks, they sized each other up—until one November evening when Hillary, slaving over books in the law library, spotted Bill talking to a fellow student in the hallway. As he listened to Jeff Gleckel try to talk him into writing for the *Yale Law Journal,* the man from Arkansas had trouble focusing. "His glance began to wander and he seemed to be looking over my shoulder," Gleckel recalled. At Hillary, as it turned out.

Finally, Hillary pushed her chair back, got up, and walked toward the two men. "Look, if you're going to keep staring at me," she said, "and I'm going to keep staring back, we should at least introduce ourselves. I'm Hillary Rodham."

Bill's mind went blank. He paused for a moment, then stuttered his name.

It would be five months before they spoke again. Hillary and David Rupert were still a couple when she decided to accept Bill's invitation to see a Mark Rothko exhibit at the Yale Art Gallery. The museum was closed because of a labor dispute, but Hillary watched in amazement as Bill talked his way in by promising to pick up litter in the museum courtyard. It was the first time, she would recall, that she "saw his persuasiveness in action." That afternoon, Hillary and the persuasive Mr. Clinton had the Yale Art Gallery all to themselves.

If, as he claimed, Bill found Hillary daunting, he didn't let on—and that alone was sufficient to impress her. The six-foot-two, 210-pound Arkansan with the Elvis drawl was, she marveled, "the one guy who wasn't afraid of me." After that first date, Hillary left to spend the weekend with David Rupert. When she returned with a cold, Bill showed up at her door with orange juice and chicken soup.

This act of unsolicited kindness marked another in a string of epiphanies for Hillary, who soon learned that Bill had harbored presidential aspirations since the age of seven. (When Billy Clinton's second-grade teacher told his mother he would be President someday, Mrs. Clinton replied, "Oh, yes—that's what I tell him every day.") Rupert was history.

Fellow lawyer and friend Terry Kirkpatrick would use one word to describe Hillary's feelings for Bill: "Besotted . . . not a word I would normally apply to Hillary, but I think she was besotted."

Not a small part of Bill's appeal was his unvarnished ambition, and Hillary's growing conviction that he would realize his boyhood dream. "He's going to make it," she told anyone who'd listen almost from the time they began dating. "He's going to change the world."

Hillary had no idea at the time that, as she put it, Bill would "cause my life to spin in directions that I could never have imagined." Nor did she fully appreciate that, no matter how affectionate he seemed toward her, Clinton was incapable of being faithful. Had she known about his twisted family history of divorce, violence, bigamy, poverty, addiction, illegitimacy, and promiscuity, Hillary might better have understood what lay in store for her.

Bill Clinton began life as William Jefferson Blythe III on August 19, 1946, and he never knew the man listed on his birth certificate as his biological father. Hard-drinking, womanizing W.J. Blythe II had gone through three wives—including a pair of sisters—before he married Virginia Cassidy, a student nurse with a fondness for garish lipstick, stiletto heels, and tight sweaters. Blythe never bothered to tell his bride that he hadn't taken the trouble to divorce his third wife. Therefore, the marriage that theoretically produced a future President was invalid.

Back home in Hope, Arkansas, Wife Number Four was guarding secrets of her own. After her husband shipped out for Europe with the 12th Battalion in 1943, Virginia returned to the wild life she had led before they married, dating old boyfriends and partying until the early morning hours.

Blythe returned from the war in December of 1945, and six weeks later Virginia announced that she was pregnant. But on May 17, 1946, Blythe was speeding down Route 61 when he blew a tire, causing his midnight blue Buick to roll over. Pulling himself from the wreckage, he collapsed in a drainage ditch—and drowned. William J. Blythe III was born three months later.

Bill's mother wasted little time landing another husband—a carousing, big-spending, loud-talking Buick dealer from neighboring Hot Springs. Virginia knew Roger Clinton had a history of violence—in court papers his previous wife described him as an unrepentant batterer—and that he was also seeing several other women. None of this mattered to fun-loving Virginia, but it did to her parents, who threatened to seek custody of their grandson if she went ahead and married Clinton. Instead they agreed to back down, in exchange for playing a major role in Billy's upbringing.

Bill's earliest memories would be of chaos—of drunken tirades and pitched battles between his mother and stepfather. At various times, Roger Clinton punched Bill's mother in the face, threw her against walls, kicked her as she lay writhing on the floor, threatened her with scissors, and fired a pistol at her. To further complicate matters, Bill was filled with despair as his doting but emotionally unstable grandmother was committed to a mental institution after suffering a stroke.

Through it all, Bill—known to family and friends as "Bubba"—somehow not only survived, but flourished. He excelled in school, and when his half brother Roger arrived in 1956, Bubba relished his role as protective older sibling. Their bond would only grow stronger as they tried to cope with the emotional turmoil swirling around them.

Bubba was fourteen when he finally confronted Roger Clinton, telling him to "never . . . ever" touch Mom again. Virginia finally divorced Clinton, only to remarry him two months after the divorce was finalized. At fifteen Bubba, always the epicenter of his mother's universe, was now the man of the house. While the living room was covered with the awards and framed certificates Billy had accumulated, Virginia—who shared a small second bedroom with her husband—willingly relinquished the master bedroom to her cherished elder son. Undeterred by the fact that his problematic stepfather refused to legally adopt him, Billy legally changed his surname to Clinton—a "gesture of family solidarity," Bubba explained, aimed at reassuring his little brother.

Roger Clinton's toxic lifestyle began to take its toll. As his health declined, he retreated to the family room, where he drank beer and stared blankly at the television set while his wife went out on the town—not always alone. On several occasions she brought sixteen-year-old Bubba along to nightclubs as her "date." (Following the death of Roger Clinton in 1967, Bill's mother would land two more husbands—hairdresser Jeff Dwire, who had served jail time for fraud and died of a sudden heart attack in 1974, and retired stockbroker Dick Kelley. Virginia remained married to Kelley until her death in 1994.)

Incredibly, none of Billy's friends or classmates knew of his hellish home life. So long as no one was aware of what was really going on behind closed doors, he could convince himself that the horrors that scarred his childhood never really happened.

Out of embarrassment, Bill Clinton was loath to share the more sordid details with Hillary. But early on, he would let her in on a secret his mother had taught him about coping. Whenever things became overwhelming, Virginia told her teenage son, "Brainwash yourself. Put the bad things out of your mind—just push them aside so they don't interfere with the important things in your life." When Bubba recoiled at the word *brainwashing,* she then came up with a visual image: "Construct an airtight box in your mind. Keep inside it what you don't want to think about. The inside is white, the outside is black. . . . This box is strong as steel."

"Boxing things off," as Hillary would later refer to the process, would free young Bill Clinton from distraction. Even the most unpleasant thoughts were neatly packaged and shelved, allowing him to concentrate on those things that mattered most to him. Once she learned of the technique, boxing off would also prove to be a godsend for Hillary. She even seemed to enjoy the game; Hillary often gift-wrapped her boxes and tied them in a bow before storing them neatly in the far reaches of a deep, deep, cedar-lined closet at the back of her mind.

Where Hillary sought to win approval from her father that was never forthcoming, Bill made up for the absence of a nurturing dad by seeking approval in the eyes of others. Toward that end, he joined every imaginable club and school organization at whites-only Hot Springs High, shamelessly sought to ingratiate himself with teachers, and ran successfully for student council. Like Hillary, Bill was congenitally clumsy and shied away from sports. Instead, he played saxophone in the marching band and, out of deference to his mother's near-fanatical obsession with Elvis Presley, broke into an eerily dead-on rendition of "Love Me Tender" whenever the mood struck him.

While Hillary was singing Barry Goldwater's praises up north, Clinton was angling to get himself sent to the American Legion's Boys Nation conclave in Washington. Once there, he pushed himself to the front of the pack so that he could get his picture taken shaking JFK's hand in the White House Rose Garden—an image that would continually reinforce the notion, in his own mind and in those of others, that he was destined for great things.

At Georgetown University, Bill Clinton's indefatigable friendliness—he made a point of meeting virtually all the two-thousand-plus undergraduates—and his down-home Southern style would quickly make him the most popular man on campus. But after serving as president of both the freshman and sophomore classes, he suffered a stinging defeat when he ran for the top office in his junior year. Paralleling Hillary's bitter senior class election defeat in high school, Bill's rejection by his Georgetown peers would leave a deep and lasting wound.

Hillary would not be surprised to learn that Bill had also been traumatized and transformed by the assassinations of Martin Luther King and Bobby Kennedy. Determined to make his own contribution, Bill spent the summer after his Georgetown graduation campaigning for Arkansas Senator J. William Fulbright, chairman of the powerful Senate Foreign Relations Committee and an outspoken critic of the Vietnam War.

As a Rhodes Scholar at Oxford, Bill also marched in several antiwar demonstrations during his time abroad, and even made a curious side trip to the Soviet Union. But one concern would overshadow all others to the point where he could not simply "box it off." Bill spent most of his time at Oxford trying to find ways to avoid being drafted—a mission that grew all the more urgent after he received his induction notice just as the American death toll crept past the forty-thousand mark.

With the help of his mentor Senator Fulbright, Bill managed to get the rules bent so he could join the Reserve Officer Training Corps. But once a draft lottery was instituted and it became clear his number was so high he would never be drafted, Bill wrote a letter to the head of the ROTC program at the University of Arkansas telling him that he was one of those who found themselves "loving their country but loathing the military." Bill had conned his way into the ROTC appointment, which he now spurned, for one reason: "to maintain my political viability within the system."

The letter, which Clinton would take pains to keep secret, would come back to haunt him decades later when he sought the presidency. Until then, not even Hillary would be aware of its existence.

By the time Hillary began dating Bill in the spring of 1971, she was just winding down her romance with David Rupert—her first and only serious relationship. Bill, on the other hand, had cut a wide swath through Arkansas, Georgetown, and the United Kingdom.

Hillary, warned from the outset that Bill was already juggling a number of girlfriends at Yale, seemed not to care. She was happiest driving around New Haven with Bill in his hideously orange Opel station wagon.

Before long, Hillary was spending weekends at the beach house in Milford, Connecticut, that he shared with three roommates. They, like nearly everyone else at Yale, were intimidated by Hillary's intellect—and by her directness. She respected Bill, and believed without question that he would someday be President. But that didn't stop her from delivering the coup de grâce whenever Bill was getting too full of himself. "Knock it off, Clinton," she'd say when he'd gone on too long extolling the virtues of his home state. "Cut the crap!"

At the beach house one night, Bill and Hillary were sitting in the kitchen, talking about their plans. Hillary was not quite sure which direction she was going in, or even where she wanted to live. She was impressed that, hokum aside, Bill loved Arkansas and knew that he wanted to hold office there.

That summer, Hillary accepted an offer to clerk for a law firm in California, and to her surprise, Bill wanted to tag along. They shared a small apartment not far from the Berkeley campus of the University of California, and while Bill spent most of his time reading and sightseeing, Hillary did research and wrote legal motions for the Oakland firm of Treuhaft, Walker, and Burnstein.

Hillary had first met Robert Treuhaft and his wife, Jessica Mitford, when they came to New Haven to raise money for the Black Panthers. More recently, she caught their eye when the *Yale Review*—Hillary was now on the *Review*'s editorial board—ran an article defending Black Panther defendant Lonnie McLucas. The piece was illustrated with drawings of policemen as pigs; one had been decapitated.

Treuhaft and Mitford, who had written a bestselling exposé of the funeral industry called *The American Way of Death,* were avowed Stalinists. Treuhaft formally resigned from the Communist Party in 1958 because by then it had lost so many members it lacked any real clout. But he and his wife, who dismissed the heroic 1956 Hungarian uprising against Communist rule as the work of "grasping neo-Fascists," remained staunchly committed to the cause. In addition to the Panthers, Oakland's "Red Lawyer" represented a wide range of radicals and indigents. In later years

Hillary would say nothing of her friendship with Treuhaft and Mitford, and refer to Treuhaft, Walker, and Burnstein only once in her memoirs, simply as "a small law firm in Oakland, California."

Hillary returned to New Haven hating the war and Richard Nixon ("He's pure evil") more than ever. She and Bill rented a ground-floor apartment just off campus at 21 Edgewood Avenue for seventy-five dollars a month, and Bill used his own money to set up a "McGovern for President" headquarters. Much of the time, Bill worried aloud that his new relationship was not fair to Hillary. He told her that he wasn't sure he wanted to fall deeper in love with her because she would never be happy in Arkansas. "If you wanted to run for office, you could get elected. You could even be a senator," he said, "but I've got to go home."

After Christmas, Bill drove up to Park Ridge to spend time with the Rodhams. While he had no difficulty winning over Dorothy and the boys, Dad was his usual implacable self. Gradually, over football and card games, Hugh Rodham began to crack. Interestingly, it was Dorothy who confronted Bill on the question of her daughter's future—and why he felt it was fair for Hillary to relinquish her own political ambitions while he sought office back home in Arkansas.

That summer of 1972, Hillary joined Bill in Austin, Texas, where they both worked on the McGovern campaign. She had managed to keep an eye on her boyfriend's extracurricular activities in New Haven. But with Bill making frequent out-of-town trips to whip up support for his antiwar candidate, there was little she could do to rein him in. At numerous times in front of other campaign staffers, they quarreled bitterly about the other women he was seeing—as many as three in the span of a single week.

Bill continued to impress everyone he met with his prodigious energy—he managed to get by on just five hours of sleep a night—and his innate political savvy. But Hillary was also attracting notice from seasoned party operatives. One of these, no-nonsense, chain-smoking Betsey Wright, was far more interested in Hillary's future than she was in Bill's. "I was obsessed with how far Hillary might go," Wright said, "with her mixture of brilliance, ambition, and self-assuredness." She would later say that Bill and Hillary's marriage would leave her feeling "disappointed. I had images in my mind that she could be the first woman president." None of this kept Hillary from signing on for an additional—and unnecessary—fourth year at Yale Law School simply so she could be close to Bill.

After graduation, Bill took Hillary on her first trip to Europe. It was while strolling along the shores of Lake Ennerdale in England's scenic Lake District that Bill first asked Hillary to marry him. With the specter of her grandparents' divorce and the havoc it caused in her own mother's life still looming large in Hillary's mind, she turned him down. "No, not now," she said. "Give me time."

Undaunted, Bill took her home to Arkansas—but not before using his contacts to secure a $25,000-a-year job teaching law at the University of Arkansas in Fayetteville. In Hot Springs, Hillary met an icy reception from Virginia and Bill's brother, Roger. "I didn't know what to think," Hillary's future mother-in-law recalled of that first meeting. "No makeup. Coke-bottle glasses. Brown hair with no apparent style."

And style—if not taste—was something Virginia had in abundance. Her skin dangerously dark from too much sun, Bill's mother favored shorts, open-toed shoes, hot pink lipstick ("I always figure the brighter the better"), and Minnie Mouse eyelashes. Her own hairstyle was distinctive all right: a bouffant dyed black on two sides, with a wide white "skunk stripe" down the middle. Virginia had wanted her son to marry someone she felt was more his type—perhaps one of the tall blond beauty-pageant contestants he seemed to favor. Hillary was equally taken aback. She and Virginia seemed to be, Hillary later said, "from different planets."

Hillary had also made several visits over the previous year to New York, in part to investigate the living conditions of disadvantaged children for her various Marian Wright Edelman–inspired projects, but also to connect with various activist groups in the city. Through her many left-wing friends and contacts—including avowed Communists Robert Treuhaft and Jessica Mitford, Black Panther lawyer Charles Garry, and her old mentor Saul Alinsky—Hillary was introduced to dozens of like-minded radicals. Among her new friends: New York–based representatives of the Palestine Liberation Organization who were lobbying feverishly to be recognized as a diplomatic entity by the United Nations. (A year later, the PLO would be granted "permanent observer" status and, ironically, would ultimately open its mission in an Upper East Side town house one block from the homes of David Rockefeller and Richard Nixon.)

At a time when elements of the American Left embraced the Palestinian cause and condemned Israel, Hillary was telling friends that she was "sympathetic" to the terrorist organization and admired its flamboyant leader, Yasser Arafat. When Arafat made his famous appearance before the UN General Assembly in November 1974 wearing his revolutionary uniform and his holster on his hip, Bill "was outraged like everybody else," said a Yale Law School classmate. But not Hillary, who tried to convince Bill that Arafat was a "freedom fighter" trying to free his people from their Israeli "oppressors."

Antipathy toward Israel was so strong on the American Left that even New York Congresswoman Bella Abzug, an active Zionist in her college days, was branding Israel an "imperialist aggressor" by the late 1960s. "It was de rigueur to support the PLO in the mid-seventies," recalled a onetime Arafat sympathizer who would later become an executive with a major advertising firm. "It was just more of the radical chic thing. But this time it was the PLO instead of the Panthers." Four years after Arafat's scorching attack on Israel at the UN, one of his staunchest supporters, Vanessa Redgrave, showed up at the Academy Awards to accept a best supporting actress Oscar for *Julia* and blast the "Zionist hoodlums" who picketed her presence at the ceremony. Redgrave was practically booed off the stage for her remarks, but according to a friend in Arkansas, Hillary defended the actress's stance. "Hillary's line basically was that there were two sides to the Palestinian question, and that Jews in this country had too much money and power and that they'd pretty much shut off all debate."

In late July of 1973, however, Hillary's mind was on practicing law. She and Bill made another trip to Arkansas to take the bar exam—and passed on the first try. (Hillary would not be so fortunate in Washington, D.C., where she had taken the exam twice before passing.) It was during this trip to his home state that Bill took Hillary to meet a politically well-connected friend. When they drove up to the house, Bill and Hillary noticed that a menorah—the seven-branched Hebrew candelabrum (not to be confused with the more common and subtler mezuzah)—had been affixed to the front door.

"My daddy was half Jewish," explained Bill's friend. "One day when he came to visit, my daddy placed the menorah on my door because he wanted me to be proud that we were part Jewish. And I wasn't about to say no to my daddy."

To his astonishment, as soon as Hillary saw the menorah, she refused to get out of the car. "Bill walked up to me and said she was hot and tired, but later he explained the real reason." According to the friend and another eyewitness, Bill said, "I'm sorry, but Hillary's really tight with the people in the PLO in New York. They're friends of hers, and she just doesn't feel right about the menorah."

"Do you mean to tell me," Bill's friend shot back, "that she is going to be part of Yasser Arafat and all those people?"

Bill shrugged. "Hillary really backs the PLO and doesn't like what Israel is up to," he said, looking sheepish. "Anyway, she just thinks having a menorah on your front door . . . she just doesn't like it, that's all."

After a brief stint in Boston doing fieldwork for Marian

Wright Edelman's Children Defense Fund, Hillary returned to Arkansas—and to Bill, who by now had decided to run for Congress. Watergate and talk of impeachment dominated the news. Bill was convinced that, as the Nixon administration proceeded to self-destruct, the Republican Party would inevitably be hard hit in the 1974 congressional elections.

Hillary soon learned what part she would play in the unfolding drama. In January of 1974, John Doar, the House Judiciary Committee's new chief counsel, offered both Bill and Hillary a chance to work on the impeachment inquiry staff. Bill was already planning his run for the Third Congressional District seat occupied by Republican John Hammerschmidt. But Hillary, who like most McGovernites never attempted to mask her hatred of the incumbent President, jumped at the chance. Now she would be able to play a role in bringing down the "evil" Richard Nixon—a man whose removal from office she called for long before the Watergate break-in occurred. She also had a vested interest in the outcome: an impeachment trial would make it especially hard for incumbent congressmen like John Hammerschmidt, Bill's conservative Republican opponent, to get reelected.

Hillary's main job was to look up legal precedents for impeachment—an experience she would later recall without the slightest trace of irony—but she did get to listen to several of the infamous Nixon tapes. In her daily calls to Bill, she shared some of the more damning details of the investigation in general and the tapes in particular. Bill was more confident than ever that, armed with this sensitive information, he could destroy Hammerschmidt at the polls.

Working up to twenty hours at a stretch, Hillary ate, slept, and breathed Watergate. Yet she was still obsessed with Bill. Whenever the opportunity arose, she blithely informed her coworkers that her boyfriend was going to be President someday. "She said it to me," said fellow staffer Tom Bell. "She said it to a lot of people."

Hillary had come to rely on one senior staff member in particular for advice. But Bernie Nussbaum made it clear that he felt Hillary's predictions about her boyfriend's future were inappropriate. Nussbaum was giving her a ride home one night when Hillary launched into her "Bill's going to be President someday" speech. Angered by Hillary's presumptuousness, Nussbaum told her it was "insane" to be making such comments while they were in the process of seeking ways to remove Richard Nixon from office.

Hillary, red-faced, ordered Nussbaum to stop the car. "Bernie," she shouted, slamming the car door behind her, "you are an asshole!"

Back home in Arkansas, Bill was already on the road to fulfilling Hillary's prophecy—literally. Crisscrossing the state, he charmed voters and began to build a formidable war chest for his campaign.

Despite her own demanding schedule, Hillary somehow managed to pepper Bill with a half-dozen or more calls a day. A harbinger of things to come, Hillary virtually ran the campaign over the phone. Speeches, schedules, staff assignments—all had to be run by Bill's girlfriend for her approval.

Hillary's involvement did not stop there. She was hearing rumors that Bill was spending time with a precocious eighteen-year-old volunteer, as well as with his old high school flame Dolly Kyle, who had just divorced her first husband. Hillary dispatched her father and her younger brother, Tony, down to Arkansas to "help out with the campaign." Translation: to check out those disturbing rumors about Bill and other women.

When they reported back that Bill was involved with at least a half-dozen women in Fayetteville and Little Rock, Hillary got on the phone and tore into Bill. Coming over the receiver, her blistering obscenities were clearly audible to campaign workers seated nearby.

Bill pleaded tearfully for Hillary to forgive him, and she did, grudgingly. Two weeks later she was at his side as the primary campaign wound down. "I've got to be there," she told a fellow member of the Watergate impeachment staff, "just to make sure they don't fuck it up."

On May 28, 1974—primary day—Bill and Hillary flew to Little Rock for one final TV interview. Bill, distracted by the pressures of the campaign, had forgotten that Dolly Kyle—his nickname for her was "Pretty Girl"—would be meeting them at the airport. The good-looking blonde in the clingy white summer dress was not prepared for the sight of Hillary—matted hair, thick glasses, body odor (even in the oppressive Arkansas heat Hillary abstained from wearing deodorant), unshaven legs, and all. For one fleeting moment, Kyle thought this was a practical joke, and that Bill had hired an actress to play Hillary. But then she "knew from looking at Hillary that she'd make him pay" for this awkward moment.

Bill handily won his party's nomination, but the campaign against Hammerschmidt would be an uphill battle. Hillary, meanwhile, returned to her job on the House Judiciary Committee staff and resumed feeding Bill valuable, behind-the-scenes details.

Hillary's most important assignment had been to draw up procedural rules for presentation of evidence to the House—in other words, the blueprint for prosecuting a case against the President. Among other things, Hillary argued that the constitutional definition of impeachable offenses was outmoded and should be disregarded, and that the President had no right to legal counsel in impeachment proceedings (her husband would have no fewer than seven lawyers represent him during *his* impeachment trial). Hillary, along with Doar and Nussbaum, also wanted Judiciary Committee members barred from cross-examining witnesses or disclosing any of the evidence in the case—rules that would have given Hillary and other members of the legal staff more or less complete control of the proceedings. "It would then have been a secret star chamber proceeding," said another staff member, "without the public ever really knowing what the evidence against the President was."

Judiciary Committee members rejected Hillary's ideas, but went ahead and crafted three articles of impeachment on their own. Fellow Judiciary Committee staff member William Dixon would go on to be a supporter of Clinton's. But even Dixon claimed that in her zeal to destroy Nixon, Hillary "paid no attention to the way the Constitution works in this country, the way politics works, the way Congress works, the way legal safeguards are set up."

There were others, like chief counsel to the Judiciary Committee Jerry Zeifman, who railed against the "polished and sophisticated arrogance and deceit" of Doar's top assistants—most notably Hillary Rodham. Zeifman accused her of withholding information from the committee in an effort to steer it in the direction she wanted. As lawyers go, said Zeifman, Hillary was "less than honorable."

On August 8, 1974, Richard Nixon went on television to announce that he would resign the next day rather than subject the nation to a drawn-out impeachment trial in the Senate. Hillary, whose own husband would not make the same sacrifice for his country despite calls for his resignation from members of his own party, was elated at the news of Nixon's resignation. Thirty years later, she would boast that the 1974 impeachment effort "forced a corrupt President from office and was a victory for the Constitution and our system of laws."

A world of possibilities now opened up to Hillary. Would she take a high-paying position with a top Washington law firm, accept Marian Wright Edelman's standing offer of a job at the Children's Defense Fund, or—as several of her closest friends thought she would do—make her own run for office?

But Hillary had already made up her mind. She was going to take a teaching position at the University of Arkansas—and do whatever she could to get Bill elected to Congress. Not that she was entirely taken with moving to the land of pie suppers, hog-calling contests, and june bugs the size of Buicks. Echoing the sentiments of all Hillary's friends, Sara Ehrman wanted to know, "Are you out of your mind? Why on earth would you throw away your future? You are crazy. . . ."

For emotional support, Hillary persuaded her brothers, Hughie and Tony, to move to Fayetteville and enroll at the university. "It was never in the game plan to grow up and fall in love with someone from Arkansas," Hillary conceded. Although she would often say she chose "heart over head—and that's never wrong," Hillary was in fact making a conscious change in her game plan—a calculated decision to cast her lot with the man she was convinced would someday occupy the White House.

Hillary had no intention of abandoning her own dreams. She would run for national office someday, she was confident of that. But from the vantage point of 1974, the presidency was the longest shot for her or for any woman. It was, conversely, not beyond imagining that Bill could pull it off. To help him in his quest for the White House and then share power as equals—this was an attainable goal, and one for which Hillary was willing to make some adjustments.

"I think she believed he had what it takes to be President—the charisma, the intelligence, the drive," said longtime Clinton family friend Carolyn Yeldell Staley. "But Hillary also knew that Bill had a tendency to be all over the map, and that he wasn't as good a strategist as she was." She would have to be there overseeing every detail, Hillary said, "to make sure there are no screwups."

None of which sat very well with veteran political operative Paul Fray and Fray's wife, Mary Lee, who were managing Bill's campaign. With her take-charge manner and frequently belligerent tone, Hillary instantly made enemies. She also resisted any attempts on the part of Bill's mother and Mary Lee Fray to get her to change her image. Hillary still refused to wear makeup, brighten up her wardrobe, ditch her glasses, or even shave her legs.

Now that she was in Arkansas, Hillary presented another problem to the campaign. She had lived openly with Bill while attending

Yale, but that kind of cohabitating could sink a candidate in the Bible Belt. So Bill stayed behind in his tiny bungalow in Fayetteville while she moved into a spacious contemporary lent to her by an Arkansas lawyer who had served with her on the impeachment staff.

Hillary believed that Watergate was the most formidable weapon in Bill's arsenal. Hammerschmidt had been a friend of Nixon's and one of his most outspoken defenders. Now was the time, Hillary told Bill, to "really let him have it. No one who backed Nixon through this thing should be allowed to remain in office. No one."

Perhaps. But Arkansas voters were often swayed by other, less weighty issues—like the appearance of the candidates' wives. Hillary's decidedly unfeminine look and humorless demeanor was anathema in a state where political wives were expected to hide behind thick coats of makeup and frozen smiles. Before long, there was talk among members of both parties that the Yankee Bill Clinton had chosen to marry was gay.

"The lesbian rumors were really starting to hurt us," said Fray, who felt he had no choice but to ask Bill if they were true. Incredibly, Bill would not deny them. Instead, he merely shrugged.

When Fray asked Hillary, she fired back that it was "nobody's goddamn business." When Fray pointed out that the gossip was losing Bill votes, Hillary blasted him again. "Fuck this shit!" she screamed before turning to leave, slamming the door of the campaign office behind her.

Hillary was proving to be a handful for other reasons as well. The Frays were expending a considerable amount of time and energy trying to conceal Bill's numerous infidelities from Hillary. The list of "Special Friends" kept by Mary Lee swelled to include scores of names—many the wives and daughters of some of his most ardent supporters. Hillary periodically vented her frustration by raiding Bill's desk, searching for his girlfriends' phone numbers, then tearing them up in a frenzy when she found them.

It was not atypical behavior for either Hillary or Bill. Both had hair-trigger tempers, and neither seemed to care if there was an audience on hand or not. Campaign organizer Ron Addington remembered pitched battles between the two—not over Bill's flagrant womanizing ("He had a woman in every county, and there were thirty counties"), but over differences in how to run the campaign.

"Politics, not sex, is what really got Hillary worked up," Paul Fray said. During one argument at campaign headquarters, Hillary picked up a book and flung it at Bill, catching him in the ribs. "She was frightening—she liked to yell, and man, did she like to throw things," Fray recalled. "Hillary could just scare the living shit out of you if she wanted to, and I mean that." Ironically, Fray and his wife were sucked into the same argument and wound up hurling projectiles and invective until Mary Fray detonated a megaton bomb. They had been given the onerous task, she told Hillary, of hiding the fact that Bill was "sleeping with half the district."

Hillary did not react with histrionics; she barely seemed to react at all. Her response led the Frays to conclude that she had been well aware of the nature of Bill's extracurricular activities, if not the scope. Still, in the aftermath of her explosive statement, Mary Fray waited for Hillary to say *something.* When she didn't, it became instantly clear to the Frays that what mattered most to Hillary right now was winning this election.

What most upset Hillary during the 1974 congressional campaign was Bill's reluctance to bludgeon Hammerschmidt over his long-standing allegiance to Nixon. Campaign worker Ron Addington was driving Bill and Hillary to another stop when they began shouting at each other in the car.

Screaming from the backseat that he was going to lose the election because he was "chickenshit," Hillary demanded to be let out of the car. Bill, yelling back at the top of his lungs, was happy to oblige. Hillary got out, slammed the car door, and stomped off alone. "You're going to lose, asshole," she shrieked over her shoulder. "You're going to lose!"

Hillary was right. Bill lost by a hairbreadth—a margin of just 1.5 percent. She was disappointed that Bill had not followed her advice, but not upset enough to keep him from moving in with her. Now both Bill and Hillary—infamous among their friends for never picking up the check—lived rent-free in Terry Kirkpatrick's large home. "Bill sponged off people," Fray said of Clinton, and Hillary "wasn't much better."

Hillary and Bill returned to their teaching jobs at Fayetteville, where she outshone him as a professor. "Bill wanted to be treated like one of the guys," said Woody Bassett, a student who had them both as teachers. "He was laid-back in the extreme." Hillary, on the other hand, "was no-nonsense, analytical, demanding—all business, all the time. People loved him, but they respected her."

Hillary also founded the law school's first legal aid clinic. In one case she was asked to represent a woman who was about to be sent to a state mental hospital for loudly preaching the Gospel on street corners. By way of a compromise, Hillary suggested that the judge buy her client a one-way bus ticket back to her home state of California, where, Hillary said, the citizens were much more in need of spiritual guidance. The questionable ethics involved in shipping the mentally ill of one state to another did not bother Hillary, who would boast of this Solomonic moment in her memoirs.

In addition to her satisfying legal work, Hillary had expanded her own circle of friends beyond Bill's legion of fans. Diane Kincaid, a political science professor at the University of Arkansas, would become Hillary's best friend. Another transplant from Washington, D.C., Kincaid shared Hillary's feminist views and debated conservative activist Phyllis Schlafly in front of the Arkansas General Assembly on the merits of the Equal Rights Amendment. Diane and Hillary were incensed when the Arkansas papers covered the debate in their women's section, and wrote mainly about the debaters' outfits and hairstyles.

Kincaid would later marry Jim Blair, chief counsel for Spring-dale, Arkansas–based Tyson Foods and one of Bill's most enthusiastic boosters. It was Blair who brought company chairman Don "Big Daddy" Tyson into the Clinton fold, and acted as go-between whenever Bill needed large infusions of cash. The fact that the flamboyant, tough-talking Tyson was both notoriously antiunion and one of the state's most egregious polluters—his poultry processing plants spewed tons of chicken feces into state waterways—was apparently something Hillary was willing to overlook. That is, as long as Big Daddy continued writing the big checks.

"Hillary was fortunate," Diane said, "in that Bill fully appreciated what she was giving up. He felt she could do anything she put her mind to, and it bothered him that he was asking her to make such a big sacrifice. But he also felt they were equal partners, and that whatever he achieved they would share fully in."

Hillary was settling comfortably into her new life in Arkansas, but she was still not completely convinced that she should marry Bill and stay. Bill, on the other hand, was more sure than ever that he needed Hillary—and that he needed her as his wife. While Hillary blamed his defeat on failing to exploit his foe's ties to Nixon, Bill believed he had lost because he was not perceived as a family man. Rumors about his womanizing had swirled around the campaign, and at one point a Baptist minister had publicly upbraided Clinton for "living in sin" with his Yankee girlfriend. If he was going to run for office again, he was going to do it as a married man.

That summer Hillary, still undecided, traveled north to touch base with family, friends, and contacts in Illinois, New York, Boston, and Washington. There was no consensus about what she should do, but Hillary was left with the impression that nothing anyone else could offer would be as fulfilling as her life with Bill.

When she returned to Arkansas, Hillary was picked up by Bill at the airport and driven straight to a little redbrick Victorian on California Street. She had admired the house before she left, and Bill was telling her that he had purchased it. "So now you better marry me," he said, "because I can't live in it by myself." They walked through the living room with the beamed cathedral ceilings and the dilapidated 1920s kitchen, and when they reached the bedroom Hillary knew what her answer was going to be. Bill had already bought an antique wrought-iron bed and covered it with bed linens from Wal-Mart.

Virginia cried on October 11, 1975—not because she was losing her son to a Yankee, but because his bride was insisting on holding on to her maiden name. Hillary had waited until the eve of her wedding to buy her wedding gown—a Victorian lace dress from Dillard's department store. The ceremony itself, conducted in the living room of their new house by local Methodist minister Vic Nixon ("Never thought I'd be married by Nixon!"), was a modest affair. Not so the reception, which was held at the home of State Senator Morris Henry and his wife, Ann. Hundreds of friends, family members, former classmates, colleagues, and political contributors crowded into the Henrys' backyard.

Ann Henry described Hillary as "ecstatic. If she had doubts about making her life in Arkansas they had vanished by that time. She was totally in love with him, and he with her." That moment would have been shattered had Hillary known that during the reception one of the guests came upon Bill making out with another woman in one of the bathrooms.

Eight months later, Bill scored his first election victory when he won the May 1976 primary for state attorney general. He faced no Republican opponent in the general election, so he was free to take on the task of running Jimmy Carter's presidential campaign in Arkansas. Hillary, meantime, took a leave of absence from the university to become Carter's deputy campaign director in Indiana. She quickly rubbed the local operatives the wrong way—so much so that at one point during a staff meeting, a slightly inebriated pol reached over the conference table, grabbed her by the collar, and told her to shut up. Hillary removed his hand, told the man never to touch her again, and stormed out of the meeting. She had complained that Carter's people in Indiana were doing a sloppy job, and she was apparently right—though Carter won the election, he did not carry Indiana.

Now that the Clintons were moving to the state capital, Hillary resigned from her teaching position and started job-hunting in Little Rock. Since the attorney general's salary was just $26,500, it was understood that Hillary would have to take up the slack to pay for their new house in the city's upscale Hillcrest district. Bill told Dolly Kyle, who had resumed her affair with Clinton shortly after he took office, that there was deliberate "role reversal" in their marriage. "Hillary was not interested in living hand-to-mouth, and she didn't want to be known only as Bill Clinton's wife. They each had their assignment, and it was spelled out clearly. He was to be the decorative one in the relationship, and she was to be the breadwinner."

Vince Foster had been a boyhood friend of Bill's, and hosted Bill's first big congressional campaign fund-raiser at the offices of his law firm—Little Rock's prestigious Rose Law Firm—in 1974. Hillary did not get to know him well until she was running the university's legal aid clinic. At the time Foster was the head of the bar committee that oversaw legal aid.

When Foster offered her a job at the firm, she took it, starting out as an associate at $25,000 a year. Hillary moved into an office adjacent to Foster's, and they shared a secretary. Tall, courtly, and handsome, Foster quickly became—with the exception of Bill—her closest male confidant. She also grew close to Webb Hubbell, a quintessential good ol' boy who, like Bill, could talk for hours about nothing in particular. Hubbell was also an expert on some of the more arcane aspects of Arkansas law, and with ease could rattle off case citations reaching back a century or more.

Hillary did not let her conscience get in the way of defending a canning company against a man who opened a can of pork and beans to discover a rat's derriere poking up at him. The sight was so nauseating, said the plaintiff, that he could not, among other things, bring himself to kiss his fiancée. Arguing that the rodent parts had been sterilized in the canning process and "might be considered edible" in certain parts of the world, Hillary somehow managed to convince the jury to award the man only a token amount. Hillary and Bill would often joke about her first legal victory for the Rose Law Firm in what she dubbed "The Rat's Ass Case."

Not long after, she took on another case that seemed to contradict much of what she had stood for. In her first criminal case, Hillary defended a three-hundred-pound man charged with assaulting his girlfriend. Prosecutors viewed the case as cut-and-dried—police would testify that the woman had been severely beaten—but were blindsided during the preliminary hearing when Hillary convinced the judge to drop the charges on a technicality.

Both "The Rat's Ass Case" and Hillary's successful defense of a man accused of brutally attacking the woman he lived with underscored her willingness to compromise her values—if that's what it took to be on the fast track to partner. "Money was extremely important" to Hillary, claimed Roy Drew, who later managed a number of the Clintons' investments. Betsey Wright concurred, citing Hillary's assigned role as Clinton family breadwinner. "Bill doesn't care about money," Wright said. "He would live under a bridge. . . . He just doesn't care. But Hillary did."

It didn't hurt that Hillary was the wife of a sitting state attorney general. Hillary insisted that "steps have been taken" to ensure that no conflict of interest would take place, but the fact remained that Rose represented the most powerful interests in the region—from real estate and retail to banking and manufacturing. Rose counted legislators, congressmen, state supreme court justices—not to mention a former member of the United States Senate—among its partners. Yet partner Herb Rule had to concede that, as the influential wife of a future governor and President, Hillary stood out as "a prize catch."

Hillary, for one, thought the arrangement was just fine. In 1977, President Carter rewarded Hillary for her help in Indiana with an appointment to the Legal Services Corporation (LSC), a federally funded nonprofit organization established by Congress. When she was asked during her Senate confirmation hearings whether the Rose Law Firm would recuse itself from cases involving organizations that received money from the LSC, Hillary waffled. In the end, she would not say yes.

While she continued to rack up hefty fees at her Arkansas law firm, Hillary oversaw an LSC budget that swelled from $90 million to $321 million—money that was used, among other things, to try to defeat California's tax-cutting Proposition 9, get Medicare to pay for a welfare recipient's sex-change operation, and support legal efforts in Michigan to give standing to "Black English" (Ebonics) as a separate language. In the final days of the Carter administration, the LSC would frantically dole out $260 million in taxpayer funds to various liberal causes in an effort to keep the money out of the hands of incoming Reagan appointees. The General Accounting Office would issue a report on the LSC under Hillary Rodham and conclude that "many of the people associated with it are uniquely reprehensible."

As she worked her way toward partner at the Rose Law Firm, Hillary was well aware that the new attorney general was being spotted with attractive women all around Little Rock. There was Susan McDougal, the feisty young wife of his old friend James McDougal, and statuesque television-reporter-turned-nightclub-singer Gennifer Flowers. Bill's torrid affair with Flowers—she called him "baby," he called her "Pookie"—would last a dozen years, and come very close to bringing an abrupt end to his presidential aspirations. Flowers laughed when Bill snickered about fooling "Hilla the Hun," and he complained bitterly to Dolly Kyle about being repeatedly nagged by "The Warden."

Both Kyle and Flowers marveled at how Bill, now one of Arkansas's best-known politicians, was willing to tempt fate. He made love to Dolly in a Cadillac El Dorado convertible parked on a public street, with the top down and in broad daylight. Twice. His chauffeur-driven limousine parked directly in front of Flowers's apartment complex on dozens of occasions, according to the building manager. And when Bill and Gennifer had sex on her sofa, Clinton refused to let her draw the shades.

There were other risks Bill was willing to take. According to several of the women involved with him, Clinton flatly refused to wear a condom. Flowers was also apparently somewhat cavalier about contraception, with the predictable result: in December of 1977 she discovered that she was pregnant. A few days later, she informed the baby's father—Bill Clinton—that she intended to terminate the pregnancy. He did not object, and the abortion was performed in late January of 1978. Two months later, with Hillary standing at his side, Bill held a press conference to announce that he was running for governor.

According to two longtime Arkansas friends, Hillary knew about Flowers's pregnancy—and was "devastated" by it. Indeed, they believe it is one of the reasons she seldom speaks of the pivotal 1978 governor's race—one of the Clintons' notable early triumphs—and went so far as to actually omit it from her memoirs. Yet there was another, even more compelling reason for Hillary to erase that period from her mind. Her name was Juanita Hickey Broaddrick.

A registered nurse who ran her own nursing care facility in the town of Van Buren, Juanita Hickey was one of the small army of Arkansas women who in 1978 volunteered to stuff envelopes and go door-to-door to help get Bill Clinton elected governor. Not surprisingly, the leggy blond Juanita caught the candidate's eye, and he invited her to drop into his campaign headquarters the next time she was in Little Rock. Juanita told Clinton she planned to be in Little Rock for a meeting of the American College of Nursing Home Administrators and would love to swing by.

Juanita checked into Little Rock's Camelot Hotel on April 25, 1978, with her friend Norma Kelsey. But when she called Clinton, he asked if they could meet at the hotel coffee shop instead. Then he called back and asked if they could meet in her room—the coffee shop, he explained, was crawling with reporters.

Reluctantly, Juanita agreed. No sooner did he step into the room than he grabbed Juanita around the waist, spun her around, and started kissing her. She pulled away. "No!" she stated firmly. "Please don't do that!"

Disregarding Juanita's objections, Bill kissed her again. This time, he bit down hard on her top lip. She tried to pull away, but he applied even more pressure with his teeth, drawing blood. Then he pushed her down on the bed. Still biting down on her lip, Bill, according to Juanita, raped her. Violently.

Once it was over, Bill stood up, put on his jacket, and politely said good-bye. "I will never, ever forget that moment as long as I live," Broaddrick said. "He was just so casual about the whole thing, as if, you know, this was business as usual for him."

Norma Kelsey would later confirm that she returned to the room only minutes after the attack to find her friend injured and in a state of shock. It would be two days before Juanita told her future husband, David Broaddrick, what had happened. She also described details of the assault to three friends, who along with Norma Kelsey would confirm that Juanita told them she had been raped by Bill Clinton within forty-eight hours of the event.

Fearing more than public ridicule—Juanita also worried that, as attorney general, Clinton had the power to silence her permanently—she did not report the incident to the authorities. "He clearly felt that he was above the law," she said, "and I knew from experience that he was capable of violence. I was angry, but I was also ashamed and very afraid. I didn't want to do anything to provoke him. I just wanted to get on with my life."

What Juanita did not know was that Bill returned home in a panic. Worried that his victim might go public and derail his campaign, he decided to confide—to an extent—in the one person whose opinion he most valued. According to Bill's version of events, Juanita was a Clinton groupie who had lured him up to her room under false pretenses and seduced him. He was ashamed and sorry, and begged Hillary to forgive him.

Hillary let fly with the customary expletives, though she was clearly more concerned with the potential political fallout than with the mere fact that Bill had once again cheated on her. Just as it was becoming apparent that Hillary did not view this problem as insurmountable, Bill confessed there was more . . .

Hillary was ashen once she was confronted with the whole story—or at least Bill's carefully parsed version of events. But, as the woman who had set up Arkansas's first rape crisis hotline and had championed laws protecting victims, she almost certainly recognized the confession of a rapist when she heard it.

No one else on the campaign was consulted about this new and potentially explosive problem. But campaign workers did notice a change in Hillary's demeanor around this time. She was quieter than usual—campaign workers had grown accustomed to her issuing commands in a nasal Midwestern car-alarm twang—and clearly distracted.

Hillary had always known what tack to take with the other women in her husband's life. In those rare moments when she even acknowledged their existence to friends, Hillary dismissed Bill's paramours as "sluts" and "trailer trash."

But Juanita was different. Voters might overlook a candidate's faithlessness, but allegations of rape—even if they could ultimately be disproven—would be virtually impossible to overcome before the election. Hillary, warned by her husband that this woman might have "misinterpreted" what went on between them in that hotel room, now had to defuse a situation that could end Bill's career—*their* career—before it had even begun.

There was a Clinton fund-raiser scheduled to be held in just a couple of weeks in Van Buren, at the home of Dr. Chris Wells. Calls were made to make sure an invitation had gone out to Juanita, who happened to be a friend of Wells. It had. In fact, Juanita had RSVP'd long before the incident in the Camelot Hotel, and the host was certain she would be attending.

Hillary and Bill flew into town on a corporate aircraft supplied by "Big Daddy" Tyson, and were met at the airport by a campaign volunteer assigned to drive them to Wells's home. The Clintons were unaware that their driver was also a close friend of Juanita's. The driver later told Juanita that, during the entire twenty-minute ride, Hillary and Bill talked about only one thing: Juanita and what to do about her. Hillary cautioned her husband not to talk to Juanita and to leave "the whole thing up to me." All Bill had to remember to do, she said, was to point Juanita out as soon as they arrived.

"I can't explain why I went," Juanita would later say of her decision to go ahead with plans to attend the fund-raiser less than three weeks after the brutal attack. She was, she reasoned, "stumbling around in a state of shock. You really aren't thinking clearly at a time like that. You know what's happened to you, but it's so painful and so unbelievable that you just put yourself on automatic pilot."

Juanita knew from the moment she stepped inside Wells's home that she had made a mistake. "What am I *doing* here?" she thought to herself. As soon as the Clintons arrived, it was obvious to Juanita that Bill was "doing everything he could not to make eye contact with me." Hillary was not so shy.

"Is Juanita Hickey here yet?" Hillary asked brightly. "Has anyone seen Juanita?" Bill, too skittish to come anywhere near Juanita, was of no help at all to his wife. Still, it didn't take long for Hillary to get someone to point Juanita out. Hillary made a beeline for Juanita, then cornered her and grabbed her hand forcefully enough to frighten her.

Hillary locked her eyes onto Juanita's, still holding on tightly to her hand. "I want you to know how grateful we are for all you've done for Bill," she said, *"and all you'll keep doing."*

Juanita tried to free herself from Hillary's iron grasp, but the candidate's wife did not release her grip for several seconds—enough for Hillary to feel certain that she had gotten her point across.

"She was looking me straight in the eye and I understood perfectly what she was saying," Juanita recalled. "I knew *exactly* what she meant—that I was to keep my mouth shut."

In that split second Juanita realized that Bill had confessed to Hillary, and that "she was not going to let that get in the way. At that moment I knew what Hillary was capable of doing. And I could see in her eyes that she wasn't doing it for her husband. She wasn't even doing it for them. She was doing it for Hillary Rodham."

When Hillary moved on to the other guests, Juanita retreated to a hallway, where she became physically ill. "I was so upset," she said, "all I could think of was 'Oh God, I'm going to throw up.' "

Five months later, Bill and Hillary stood before hundreds of jubilant supporters in the ballroom of the Camelot Hotel—the same hotel where he allegedly assaulted Juanita Hickey—to claim a decisive victory in the governor's race.

Back in Van Buren, one of Bill's earliest and most ardent fans—a lifelong Democrat who had been active in the party for years—was trying to take it all in on television. "It was just . . . you know, *surreal,*" said Juanita, who paid special attention to Hillary "laughing and waving to the crowd. She was just so pleased with herself. I kept thinking, 'What kind of woman is this? What kind of woman?' "

This is a woman whose future is limitless.
She could be anything she decides to be.

—*Bill Clinton*

■

Take a good look at her. She'll probably be the President of the United States someday.

—*Nancy Wanderer, Wellesley class of 1969*

It's not so much she screams—it's more the tone in her voice, the body language, the facial expressions. It's *The Wrath of Khan*.

—A colleague at the Rose Law Firm

■

Sometimes, the devil's in that woman.

—The Clintons' cook at the Arkansas Governor's Mansion

3

"Money means almost nothing to Bill Clinton," Hillary would say. It was left up to Ms. Rodham—Hillary still had no intention of taking her husband's name or softening her image—to "build up a nest egg. It was a pretty substantial burden on me personally. . . ."

But one, it seemed, she was eager to take on. Hillary intended to use her newfound status as Arkansas's First Lady to make money, and lots of it. "She had a taste for the finer things," Dick Morris observed. "She craved luxury." Even before he took the oath of office, Hillary complained that his annual salary as governor was a meager $35,000. She left out the fact that she would now be living rent-free in the stately Governor's Mansion, a two-story redbrick Georgian-style mansion set on six acres in Little Rock's historic Quapaw quarter. There were other perks as well: a small army of gardeners, maids, housekeepers, cooks, and security guards to keep things running smoothly at the mansion, access to several chauffeur-driven limousines, even a $51,000 annual food allotment. In the end, the governor and his wife would cost Arkansas taxpayers more than $750,000 a year.

This was not enough for Hillary, however. Now that her husband was governor and responsible for doling out lucrative state bond work to Arkansas's law firms, she was more valuable to Rose than ever. Gradually, she would use her influence to be appointed to a number of corporate boards—and in the process collect a small fortune in director's fees.

At various times during her decade-long stint as Arkansas's First Lady, Hillary served on the boards of Wal-Mart, the French chemical company Lafarge, the Little Rock–based national yogurt chain TCBY, and the Southern Development Bancorp. TCBY was a loyal client of Rose during Ms. Rodham's tenure on the board, paying upward of $750,000 in fees to the firm. Southern Development paid the firm an estimated $200,000 in fees.

Within a decade, Hillary would be earning a little over $180,000 annually in salary and director's fees—none of which would have to be expended on rent, food, transportation, health care, or any of the other costs that must be borne by the average citizen. But Hillary wanted more.

In the months leading up to the election, Jim and Susan McDougal approached their friends with a proposition. They asked Hillary and Bill to form a partnership with them for the purpose of buying 203 subdividable lots overlooking the White River. The property, which would cost only $202,000, was perfectly situated for anyone seeking to build a retirement or vacation home.

"Nothing worked better at putting Bill to sleep than business talk," McDougal said. "But Hillary was all ears. She asked a lot of questions, and by the time we finished explaining it all to her, boy, did she want in." The two couples formed the Whitewater Development Corporation, borrowed $20,000 for the down payment, and financed the rest. Throughout the negotiations, the Clintons never bothered to visit the land—and they never would.

The McDougals' glowing sales pitch notwithstanding, the market for the lots never materialized and the deal quickly went south. In the end, Hillary and Bill lost over $60,000 on their investment—and unwittingly began a chain of events that would threaten to bring down a President.

The ink on the Whitewater deal was scarcely dry when their old friend and Tyson corporate attorney Jim Blair approached Hillary with yet another moneymaking scheme. He claimed to have made a killing investing in cattle futures, and introduced Hillary to his longtime trader at the Refco brokerage house, Robert L. Bone.

"Red" Bone was a high-stakes gambler, and for that reason alone he may have been perfectly suited to the risky commodities game. With wildly fluctuating markets and margin calls that could bankrupt an investor in seconds, commodities trading—the buying and selling of contracts in such staples as corn, cattle, wheat, and hog bellies—was not for the faint of heart.

Eventually, Bone would be suspended for three years by market authorities for "serious and repeated violations" of trading procedures, and Refco would pay a $250,000 fine—at the time the largest in the history of the Chicago Mercantile Exchange. Certainly his client Hillary seemed to be playing outside the rules. To begin with, commodities contracts are highly leveraged. Even though the minimum purchase required just to open a trading account was $12,000, Hillary was permitted to buy ten cattle futures for only $1,000.

Hillary would later claim that, simply by reading *Barron's* and the *Wall Street Journal,* she became an overnight whiz at commodities trading. And despite her claim that she was merely following Blair's advice and riding the bull market in cattle futures, she actually sold short—bet on the market price going *down*—on her first day of trading. This was only the first of scores of sophisticated trades extending over the next nine months.

What made things a bit easier for Hillary—aside from the fact that she never put more than $1,000 on the table—is that the dreaded margin calls that wiped out other speculators during this period never seemed to apply to her. At one point, Hillary's account showed a deficit of $20,000, and if she were any other investor, she would have been subjected to a margin call of $117,500—and therefore forced to cough up $137,500. Instead, the margin call was never made, and Hillary rebounded the next day after making a phenomenal series of perfectly timed trades. Curiously, the records of two of Hillary's most profitable trades vanished.

Hillary claimed she made all her trades herself. In truth, there was no proof that she ever made the transactions that, in the opinion of most experts, could only have been made by a seasoned insider. During several of her biggest transactions, Hillary was actually presiding over meetings in Washington as chairman of the Legal Services Corporation.

In the end, Hillary left the table with $100,000—a 10,000 percent return on her investment. "I was lucky," she said with a shrug when asked to explain her spectacular good fortune. How lucky? The *Journal of Economics and Statistics* studied Hillary's trades and came to the conclusion that, without the intervention of her strategically placed pals, the odds of her pulling this off were 1 in 250 million.

Hillary stopped playing the commodities game in July 1979, at about the same time she told Bill she was pregnant. "I lost my nerve for gambling," she said, explaining that the profit she had made suddenly seemed like "real money we could use for our child's higher education." Yet, since Hillary had been insulated from the rules and virtually assured by her well-placed friends of an eye-popping return on her tiny investment, it remained to be seen how much nerve it really took.

According to Hillary, the Clintons had been trying to have a baby for two years, and were just about to talk to a fertility specialist when she conceived while vacationing in Bermuda. Hillary and

Bill were both elated at the news for more than the obvious reason: the baby would be arriving just as Bill's campaign for a second two-year term swung into high gear.

Hillary talked Bill into going to Lamaze classes with her, but beyond that made no adjustments in her hectic schedule. More intent than ever on becoming partner at her firm, she insisted on maintaining her normal caseload. The strain became evident as she approached her March due date, and Hillary's obstetrician ordered her not to accompany Bill to Washington for the annual White House dinner honoring the nation's governors. A frustrated Hillary sat at home and fumed.

When Bill returned, Hillary's water broke and she went into labor—three weeks early. She was rushed to Little Rock's Baptist Medical Center, where it was quickly determined that the baby would have to be delivered by cesarean section. Up until that point, no fathers were permitted to witness cesarean births. But Hillary and Bill were both adamant that he could handle it; Virginia was a trained nurse anesthetist, and had taken Bubba to witness a number of operations as a child.

The doctors relented, and Governor Clinton was standing by in scrubs when Chelsea Victoria Clinton arrived on February 27, 1980. Hillary and Bill had arrived at the name back in 1978, while vacationing in England during the Christmas season. They were strolling through London's Chelsea district when they heard Judy Collins's version of Joni Mitchell's "Chelsea Morning." At that moment, they both agreed that if they ever had a daughter, that's what they'd name her: Chelsea. The name they selected for a boy was less imaginative—and predictable: William.

After taking a four-month maternity leave, Hillary turned the day-to-day care of Chelsea over to a live-in nurse and returned to her office. There would be no more children; doctors warned that another pregnancy might jeopardize the lives of both mother and child.

Hillary threw herself back into her work, confident that her husband would handily defeat his Republican opponent, political newcomer Frank White—so confident, in fact, that she continued to rebuff her husband's repeated pleas that she append his name to hers. Hillary also turned a deaf ear to friends who complained that they did not like receiving printed invitations from "Governor Bill Clinton and Hillary Rodham."

As it happened, 1980 was turning out to be a challenging year for the nation as a whole. Interest rates were up, the economy was down, and the Iran Hostage Crisis became a focal point for widespread frustration over the Carter administration's seemingly ineffectual foreign policy.

Neither Hillary nor her husband was prepared for Jimmy Carter's decision to relocate twenty thousand Cuban refugees—primarily mental patients and convicts who had been shipped by Castro to the U.S. as part of the notorious Mariel boat lift—to Fort Chaffee, Arkansas, for resettlement. A political bombshell was in the making. Furious that they had been betrayed by their friend from Georgia, Hillary urged her husband to call the White House and demand that they rescind the order. When they refused, Bill flew into one of his legendary purple-veined tantrums.

Publicly, the Clintons said nothing about their tiff with the Carter administration. Hillary believed it was important that, with their sights already on the White House, Bill not appear to break ranks with the national leadership.

Just how high a price the Clintons would have to pay began to come into focus in June, when nearly a thousand young Cubans protesting their confinement broke out of the camp and swarmed toward the neighboring town of Fort Smith. Bill called out the National Guard to round up the Cubans and restore order.

A few days later, Hillary insisted on accompanying Bill as he went to confront Fort Chaffee's commander, General James "Bulldog" Drummond, and demand federal help in containing the prisoners. But, on orders from the White House, none would be forthcoming.

When the dust had settled later that summer, Hillary was confident Arkansas voters would see her husband as the man who took action and contained the violence. But the Republicans took advantage of the fact that Bill had not openly criticized his friend the President. Frank White's campaign ran a series of TV ads showing footage of rioting Cubans with the voice-over: *Bill Clinton cares more about Jimmy Carter than he does about Arkansas.*

The Clintons did not buy airtime to refute White's allegations. "The charges were so ridiculous," Hillary said, "we didn't think we had to answer them."

The Cuban problem was only one of several the thirty-four-year-old governor faced. Bill's decision to raise car license fees proved wildly unpopular. There were also those who were highly critical of the brash young governor's style. Bill's chronic lateness, the unshorn locks that tumbled over his ears and shirt collar, and the scruffy, laid-back style of his youthful staff rubbed many churchgoing voters the wrong way.

Hillary, who in the past could always be counted on to grab her husband by the lapels—literally—and make him face the political realities, had absented herself from the process. "Hillary was so involved in her work at the law firm, and in trying to make up for Bill's lousy salary, that she just didn't focus," *Arkansas Times* columnist Ernest Dumas recalled. "Neither one of them thought much of White, who'd never run for office before and just struck them as sort of an oddball."

In the final days of the campaign, Hillary finally woke up. She had been the one who convinced Bill to fire New York pollster Dick Morris after Morris had helped them win in 1978. Now she was on the phone begging Morris to rescue Bill. Unless he denounced Carter, Morris said, Bill would be going down to defeat.

Hillary was with her husband at the Camelot Hotel when

exit-poll results came in showing that Bill had been reelected by a wide margin. The Clintons were thrilled—Morris had been wrong. But as the returns began to trickle in, the incumbent and his First Lady grew more and more somber. By the end of the evening, it was clear that Clinton had been defeated by just thirty-one thousand votes.

Bill was crushed by the defeat, and spent the next several days, in Hillary's words, "wallowing" in self-pity. Springing into action, Hillary called Dick Morris in New York and again pleaded with him to help. Then she convinced her husband to call their old friend Betsey Wright in Washington and persuade her to help him relaunch his career. Just ten days after his defeat, Wright, who still felt that Hillary was the more impressive politician of the two, agreed to take on the challenge.

Bill signed on with a Little Rock law firm, Wright, Lindsey and Jennings, and the Clintons moved into a yellow Victorian house in their old, upscale Hillcrest neighborhood. At Hillary's urging, however, he devoted himself essentially full-time to recapturing the state house.

Toward that end, Clinton made it his personal mission to apologize to nearly every voter he encountered—whether it was at church, on the street, in the supermarket checkout line, or over scrambled eggs at a local coffee shop—for "losing your trust."

Hillary grudgingly accepted the need for this mea culpa–thon, though she found it more than a little demeaning. "If that's what people want to hear," she acknowledged with a shrug, "then by all means give it to 'em."

Bill went so far as to invite local clergymen to come to the mansion prior to his departure and pray for him—a tactic he would later employ to help save his presidency. He did not always wear a hair shirt, however. Occasionally he'd pick up the phone in the middle of the night and blast a member of the press for, as he put it, "fucking me over." Suffering from the combined effects of rage alternating with abject contrition, Bill became increasingly depressed—and Hillary became more and more concerned.

Dick Morris remembered that, during this period, Bill Clinton was "walking around in a daze, shell-shocked—a very sad, confused man." When he wasn't apologizing to friends and total strangers—or spewing invective at hapless reporters over the phone—Bill was spending more and more time with Flowers, Kyle, and a half-dozen other women stashed around Little Rock.

At their new home, life became a series of knock-down, drag-out screaming matches. Bill wanted Hillary to comfort him, and Hillary demanded that he stop feeling sorry for himself. "Get off your ass," she yelled at him, "and do what you have to do to beat Frank White."

Hillary was also willing to do what she had to do to get back into the Governor's Mansion. When the Clintons held a press conference on Chelsea's second birthday to announce Bill's candidacy, it also marked the unveiling of Hillary's new look. Gone were the thick glasses, replaced by soft contact lenses. Her dull brown hair, which had pretty much had a life of its own, was now styled in a flip and dyed a honey blond. Gone were the baggy jeans and the shapeless peasant dresses, replaced by form-flattering skirts and blouses in a variety of pastel shades. And, while she was still registered to vote under the name Rodham, Hillary declared that henceforth she would be known as "Mrs. Bill Clinton." Hillary explained that she felt it was "more important for Bill to be Governor again than for me to keep my maiden name."

There were other, covert steps Hillary took to ensure Bill's election to a second term. She worried about the rumors of rampant infidelity—in part because she was hurt by them personally, but mostly because she knew stories like these could derail Bill's chances of a political comeback. In early 1982, she contacted a private investigator named Ivan Duda to compile a list of the women Bill had allegedly been seeing. One of the eight women

Duda identified was, as it turned out, on the staff of the Rose Law Firm.

Duda, who said he met once with Hillary in a restaurant parking lot but communicated with her for the most part by phone, was surprised at how coolly she responded when he read the names off to her. Clearly, the purpose of Duda's investigation was damage control—to be prepared for any allegations that might be aimed at them during the next campaign.

The main goal had been to project a candidate who was more mature and less arrogant. Toward that end, Bill taped a television spot apologizing to the people of Arkansas for raising car license fees. Hillary did her part to convince voters she wasn't really the pushy Yankee feminist they had been led to believe she was. Going door to door with Chelsea perched on her hip, Hillary now spoke with a faint twang and peppered her conversation with *y'alls*. As soon as Hillary was back home and away from the press and the voters, Chelsea was handed over to the full-time nanny the Clintons had brought with them when they moved out of the Governor's Mansion.

Behind the scenes, Hillary cracked the whip whenever Bill stepped out of line. After she learned that her husband had gone out drinking with aides following a campaign swing through the northern part of the state, Hillary fired off a blistering stream of obscenities. While staff members cringed, she reminded Bill that their teetotaling opponent was having a field day with rumors of substance abuse in the Clinton camp. "His face was turning bright pink," said one of those present, "and Hillary was leaning into him and screaming. The message was: Hillary was sick and tired of saving his ass every time he fucked up, and he better start taking this shit seriously."

With Hillary dogging his every step as well as playing the role of wife and mother to the hilt, Bill shaped up enough to win. The Hillary who walked back into the Governor's Mansion in

1983 bore scant resemblance to the Hillary of old. The contacts, the new, more flattering hairstyle, and the suburban wife wardrobe all but erased the memory of the ill-kempt campus radical–turned–First Lady of Arkansas.

Inside, of course, nothing had changed. Hillary was more committed than ever to making her mark as the first woman partner at Rose Law Firm, and to making as much money as the law would allow. She also continued to see herself and her husband as the joint architects of social change. During Bill's first term, Hillary had asked to be put in charge of health care reform, and as chair of the Rural Health Committee spearheaded efforts to recruit more doctors, nurses, and other health care professionals into remote parts of the state where little or no medical care was available.

Bill, who continued to praise Hillary as "the smartest person I've ever known," had no qualms about his wife's ability to handle the most daunting assignments. Nor did Hillary. Besides, The Plan, which required that Hillary would hold off on pursuing her political dreams until Bill had realized his, also carried with it the explicit understanding that power would always be shared—regardless of who actually held office.

Now Hillary asked her husband to give her another high-profile assignment—one worthy of a "co-governor." She wanted to take on the task of spearheading the court-ordered reform of Arkansas's abysmal public education system, which virtually tied with Mississippi as the worst in the nation. Hillary immediately determined that a full 1 percent hike in the sales tax would be needed to finance the $185 million in improvements she had in mind, though selling any tax increase to the voters was going to be tough. The Clintons would have at least to give the appearance of standing up to that powerful ally of the Democratic Party, the teachers' union. By calling for mandatory testing of teachers, Hillary would distract the electorate from the hefty boost in taxes.

With the hard decisions made and a strategy in place, Hillary struck out on one of the "listening tours" for which she would become famous. Visiting all the state's seventy-five counties, Hillary conducted marathon committee meetings in high school gyms, living rooms, and town halls. She invited educators and corporate executives alike to air their views over tea at the Governor's Mansion—all designed to create the impression that Hillary cared about what they had to say.

When she finally did spring the mandatory teacher testing proposal on the public, the union reacted according to plan—with outrage. Hillary was booed and jeered by teachers, but she stood her ground, arguing that it was time to weed out teachers who did not make the grade.

Hillary's call for accountability resonated with the vast majority of Arkansans, but not all the state's legislators were willing to go along with the plan. Hillary took it upon herself to call up lawmakers and cajole, persuade, or harangue them until they fell into line.

It would become evident in later months that Hillary's so-called hard-line stand against the teachers' union was, in the words of one Little Rock politico, "just for show." As it turned out, the testing system was so lax—teachers would be allowed to take remedial training and then be tested and retested over the years until they passed—that less than 3 percent of Arkansas teachers would be removed from the classroom.

Meantime, Hillary's plan forced schools to adopt strict state education guidelines—including a new emphasis on "multicultural" studies at the expense of traditional American history courses—or be shut down. The new state-mandated rules were not only costly—many districts had to raise local taxes just to keep up—but time-consuming for administrators. Sitting atop the Everest of paperwork was Hillary, who now ruled unofficially from the Governor's Mansion as the state's new education czarina.

Over the years, Hillary would repeatedly boast of her courage in holding teachers accountable. She would fail to point out that, as a practical matter, teachers were not held accountable, and that the result was a bloated state bureaucracy. What of the children? In 1986—the first year Hillary's mandated changes were fully in effect—American College Test (ACT) scores actually *declined*. As of 2004, in terms of academic performance Arkansas's children still remained at or near the bottom of the pile.

"Hillary saw herself as co-governor," said the Clintons' longtime friend Guy Campbell, "and hell, for all intents and purposes, she was." That, as well as wife, mother, law partner, guardian of her husband's political future, and major breadwinner.

It was in the role of political guardian that Hillary stepped in when Bill was informed that police suspected his aspiring rock-singer brother of dealing cocaine. Roger Clinton, who had been using cocaine since the 1970s and now sold drugs to support his four-gram-a-day habit, was videotaped as part of a sting operation. An arrest, drug enforcement authorities told the Governor, was imminent.

Bill's first inclination was to intervene, but Hillary refused to let him. She told him that if he interfered in any way, their enemies would use it against them in their upcoming bid for a third term. "And we'll lose everything," she warned him.

Bill let the investigation run its course, and Roger was arrested in August of 1984. Both Roger and their mother were furious that Bill had not warned his brother, until Hillary explained the political ramifications to them. She also took advantage of the situation to persuade all the Clintons to get family counseling. Hillary took the lead during these sessions, suggesting that Roger Clinton Sr.'s alcohol-fueled rages against their mother had permanently scarred both Clinton boys. Hillary would later say that her husband finally faced up to the fact that his stepfather's alcoholism "and the denial and secrecy that it spawned" would "take years to sort out."

For all the soul-searching, Hillary was the first to realize that

Roger's arrest could only redound to her husband's benefit. Arkansas voters, many of whose family's had been touched by the drug problem, could sympathize with the pain the governor was going through. Moreover, by not interfering in the investigation, Bill proved that he was a man of principles, unwilling to use his influence even to save his own brother. It came as no surprise to Hillary when they were swept back into office in 1984 by a margin of two to one.

There were other, potentially more dicey ramifications stemming from Roger's arrest. Roger was sentenced to two years at the Fort Worth, Texas, federal prison, but only after agreeing to testify for the prosecution in other drug cases. One of those who went to jail because of Roger's testimony was Dan Lasater, a flamboyant racehorse owner and investment banker who contributed heavily to all of the Clintons' campaigns. When it came to drugs, Lasater flouted the law, openly serving cocaine at parties attended by some of the state's wealthiest and most influential citizens—including Hillary and Bill. While she worried that Lasater was a bad influence on her husband—several eyewitnesses would step forward over the years to describe Bill's own marijuana and cocaine use—she nonetheless went along for the ride when Lasater offered to fly the Clintons to the 1983 Kentucky Derby as his guests.

For Lasater, the relationship with Bill and Hillary proved lucrative indeed. Under the Clintons' watch, his bond trading firm, Collins Locke & Lasater, was chosen to underwrite $637 million worth of state bond offerings—for a grand total of $1.7 million in fees.

It was not the last favor they would do for Lasater. When the First American Savings and Loan Association of Oak Brook, Illinois, went under, the Federal Deposit Insurance Corporation sued Lasater and his firm for $3.3 million. The Rose Law Firm represented the FDIC in the action, and Hillary—who never divulged her relationship with Lasater to the FDIC—worked out a settlement that let him off the hook for a mere $200,000.

The governor, meantime, pursued his favorite hobby with a vengeance. His confidence restored after the 1983 election victory, Bill had embarked on what amounted to a nine-year-long sexual binge. Several state troopers who had been assigned to protect the governor would later testify under oath that they had been used as virtual procurers, making it possible for him to—by his own tally—have sex with "hundreds of women."

In sworn affidavits, troopers Larry Patterson, Larry Douglass Brown, and Roger Perry recalled approaching several women *per week* for their boss. Whenever someone caught the governor's eye—whether it was at a reception or merely standing on a street corner—the troopers would be expected to contact the woman, get her phone number, and arrange for an assignation.

Sometimes the trysting place was the Governor's Mansion itself. Patterson and Perry claimed that they sneaked women into the residence while Hillary and Chelsea slept upstairs, then stood guard at the bottom of the staircase to warn Bill if his wife woke up for any reason. "Hillary Watch," they called it. At other times, Clinton would pick up someone at a bar or nightclub and drive back to the mansion, where they would park in the driveway and have oral sex in the back of the governor's official limousine.

While it is doubtful that she would have tolerated such behavior in such close proximity to their daughter—at one point Bill actually had late-night sex in the parking lot of Chelsea's elementary school—she was not entirely in the dark. She not only knew several of Bill's paramours, but was friendly with some of them. Jim McDougal was convinced his wife Susan was having an affair with Bill, and though she denied it, Bill confided in L. D. Brown that they were lovers. Years later, when he was subpoenaed to testify in a Whitewater-related case, Bill told Dick Morris he didn't know how to respond if he was asked whether he had sex with Susan McDougal. Fortunately for Bill, no one asked.

Bill also indulged his passion for beauty queens. In 1983, he began an affair with former Miss Arkansas Sally Perdue, who claimed that during one of their trysts he climbed out of bed and serenaded her with his saxophone—clad only in one of her black negligees. During his bid for a third term the following year, the governor set the rumor mill into overdrive as he pursued former Miss Arkansas Lencola Sullivan. Hillary chose to dismiss the stories as false, but she agreed with her husband's advisers that talk of such an interracial romance would sabotage his chances for reelection. Sullivan packed up and left the state.

Hillary had no idea at the time that her husband was also in hot pursuit of Miss America 1981, Bonneville, Arkansas, native Elizabeth Ward Gracen. The former Miss Arkansas met Bill at a benefit in 1983, and he gave her a lift back to her apartment in his limousine. According to Gracen, several days later they had sex in her apartment—rough sex, during which he bit down on her lip and caused it to bleed. The painful encounter, so reminiscent of his attack on Juanita Broaddrick, would leave Gracen feeling frightened and confused. Before Hillary had a chance to discover that her husband had once again placed them both in political jeopardy, Gracen flew off to New York. Bill called her repeatedly, pleading to take up where they'd left off, but Gracen held firm.

That same year, Hillary would learn of yet another affair that had the potential of sinking her husband's career—one that would be grist for Joe Klein's bestselling roman à clef, *Primary Colors.* In late 1983, Bill was jogging along his regular route near the Governor's Mansion when he encountered twenty-four-year-old Bobbie Ann Williams, one of the young black women working the stretch of Spring Street known as "Hookers' Row." She claimed that she had sex with him on thirteen separate occasions over the next several months, including one evening when she brought along two other prostitutes to fulfill Bill's ménage à trois fantasies.

Williams told a tabloid that Bill "just laughed" when she told him she was pregnant with his baby. And so did Williams's own family—

until she gave birth to Danny Williams in 1985. Williams's son was white, and with each passing year his resemblance to Clinton grew stronger. After Bobbie Williams was imprisoned on prostitution and drug charges, Danny went to live with her sister, Lucille Bolton, in one of Little Rock's poorest neighborhoods.

When the boy was three, a local activist and self-styled provocateur named Robert "Say" McIntosh began distributing pamphlets claiming Danny was Bill Clinton's "love child." Upset by the publicity and hoping to strike a deal with the Clintons, Bolton called the Governor's Mansion and managed to get through to Hillary.

Bolton had expected Hillary to sound upset, but instead she was calm, businesslike. "Is it true," Hillary asked almost matter-of-factly, "that he has this illegitimate child?" When Bolton told her the stories were true, Hillary put her in touch with a private security company that specialized in squelching such talk. "Don't worry," she told Bolton. "These people know how to stop rumors."

But they didn't. A few months later, Bolton returned with Danny. This time, Hillary refused to see her. At that point Bolton, furious that Hillary and Bill would let the governor's son grow up in soul-crushing poverty, began a campaign of her own. Periodically, Bill or Hillary would look out into an audience and see someone holding a sign demanding that Clinton provide JUST ONE DROP OF BLOOD to prove paternity.

Bill simply shrugged off the rumors, but Hillary recognized them for what they were—a growing threat to his presidential prospects. During one meeting with party leaders in Chicago, Clinton angrily denied that there was any validity to the stories. So why not simply provide a blood sample and put the rumors to rest? someone asked. Bill shook his head and changed the subject. At no point, in fact, would Bill willingly provide the blood sample that supposedly would have established that he was not Danny's biological father.

While they viewed Bobbie Williams and Lucille Bolton as little more than a nuisance, the Clintons regarded McIntosh, who had a talent for making headlines, as more of a threat. In staff meetings, Hillary grew increasingly impatient with her husband's unwillingness to do anything about the rumors. "This is dangerous, Bill," she told him. "People are starting to believe this crap. We've got to do *something.*"

In exchange for no longer championing Danny Williams's cause, McIntosh hinted that Hillary had promised him $25,000—an amount he would later sue to collect. He also claimed that Bill had promised to shorten the fifty-year prison sentence of his son Tommy McIntosh, who had been convicted of cocaine distribution. The other shoe would drop on January 20, 1993—the day Bill Clinton was sworn in as President. It was also the day that Arkansas Acting Governor Jerry Jewell signed pardon papers for Tommy McIntosh that had been prepared by Clinton before he left Little Rock for Washington. "If there was no deal, how did this happen?" Say McIntosh later asked. "How did my son get out of prison eighteen years before he was eligible for parole?"

The *Star* would announce that it had finally settled the issue in January 1999, when it compared a blood sample from Danny Williams with the profile of Bill Clinton's DNA made public by Independent Counsel Kenneth Starr. The tabloid claimed the data proved Danny was not Bill's son—a finding other DNA experts contested on the grounds that there was insufficient information in the Starr Report to make any valid comparison.

Even as tongues wagged about this and other gubernatorial peccadilloes, the mask of domestic equanimity never slipped. Hillary gazed at her husband adoringly when he spoke, and praised him for his vision and leadership whenever the opportunity presented itself. In return, Bill acknowledged her brilliant legal mind, her contributions toward health and educational reform in the state, and—most important to Arkansas voters—her parenting skills.

Privately, the picture was not so pretty. Hurt and humiliated,

Hillary frequently lashed out at her husband within earshot of the staffers. Once, after returning from one of his nocturnal expeditions in the early morning hours, Bill was surprised to find Hillary waiting for him in the kitchen. With several staff members standing just outside the door, the Clintons shrieked at each other to the accompaniment of shattering glass and slamming drawers. When it was over, staff members cautiously pushed open the door to reveal broken glass, smashed dishes, and a cupboard door ripped off its hinges.

To be sure, Hillary had continued her habit of hurling objects at her husband—yellow legal pads, files, briefing books, car keys, Styrofoam coffee cups—often in the presence of the governor's aides. Pitched battles—always instigated by Hillary, said eyewitnesses—frequently occurred in the governor's limousine. "They'd be screaming at each other, real blue-in-the-face stuff," one of their drivers said, "but when the car pulled up to their destination it was all smiles and waving for the crowd." Other times, they would sit in the car for as long as two hours without ever uttering a word.

Hillary made it abundantly clear to her husband that, while he pursued other women, her own needs were not being met. Trooper Roger Perry, a member of the governor's security detail, saw Bill virtually every day for seven years. One Sunday afternoon, Perry was standing next to an intercom outside the kitchen when he clearly heard Hillary tell her husband, "Look, Bill, I need to be fucked more than twice a year."

At one point, Bill's cheating pushed Hillary over the edge. After learning that her husband had led the daughter of a major contributor to believe he would marry her, Hillary had what some described as a nervous breakdown. She began hyperventilating, and an ambulance rushed Hillary to the hospital for observation.

Understandably upset over Bill's egregious philandering—and concerned about how it could derail their plans for conquering the White House—Hillary took out her frustrations on the governor's partners in crime: the troopers. She called them "shit-kickers," "rednecks," "hicks," and "white trash," and ridiculed them for being overweight. She also resented their constant presence and the loss of privacy that entailed. At times, a simple "Good morning, Mrs. Clinton" could provoke an attack. "Fuck off!" she would bark. "It's enough that I have to see you shit-kickers every day. I'm not going to *talk* to you, too. Just do your goddamn job and keep your mouth shut." She went so far as to instruct Trooper Patterson not to utter a word when they went out in public. "You sound," she explained contemptuously, "like a hick."

Hillary felt much the same way about Arkansans in general, even though she did a masterful job of concealing her contempt from the general public. L. D. Brown remembered driving Hillary to "the state fair, and there she was chatting up the guys in their bib overalls and the ladies in their gingham dresses—good, decent people, you know?—and they were just thrilled out of this world to meet her. Then Hillary would get in the car and say, 'My God, did you *see* that guy? He was like something out of *Deliverance*! Get me the hell out of here.'"

Like members of her husband's senior staff, the troopers were for the most part terrified of Hillary and took pains not to cross her. Only veteran officer Ralph Parker was willing to risk invoking her wrath. When Hillary received an award as Mother of the Year, Parker and other members of the Clinton entourage waited outside the Governor's Conference Room where the ceremony was about to take place. "'Mother of the Year'?" sneered Parker, who knew how little time Hillary had actually spent with her only child. "How about '*Motherfucker* of the Year?'" The rest of the staff, said one eyewitness, "looked as if they had been struck by lightning. They were scared shitless that Hillary might have heard. It was a truly great moment."

Hillary had more than her husband's constituents, her security detail, and Bill's womanizing to contend with. Several of her law partners were complaining that Hillary's billings had slipped, that she wasn't living up to her initial promise as a big earner for the firm. Worried that she might be ousted from Rose Law if she did not perform up to expectations, Hillary asked her Whitewater Development partners Jim and Susan McDougal to send some business her way—namely, she asked to be put on retainer as the counsel for the thrift they owned, Madison Guaranty Savings and Loan. In that capacity she would, among other things, represent the savings and loan in its dealing with the Clinton-appointed state securities commissioner. It would be years before Hillary's billing records, which mysteriously disappeared as federal regulators were closing in, just as mysteriously materialized in her White House office—affirming that she hid her involvement from the FDIC and other agencies.

What Hillary may or may not have known was that the McDougals engineered a series of fraudulent real estate deals to siphon off $17 million for themselves and a few select friends. Madison Guaranty finally collapsed in 1989, triggering the investigation that would ultimately lead to the McDougals' conviction on charges of mail fraud and conspiracy.

For the first time, Hillary came under heavy attack for these and other conflicts of interest during the 1986 election campaign. There were also rumblings that former Governor Frank White, who had again been chosen by the Republicans to unseat the Clintons, would for the first time make an issue of Bill's philandering.

Now that Chelsea was six, Hillary worried that she might be traumatized by the things that were being said about her parents. Chelsea had always been, according to family friend Carol Staley, a "precocious child—perfect manners, with a vocabulary far beyond her years. Her parents were away a lot, so when they were around she was eager to please them."

Hillary had always been able to control what Chelsea saw or heard about her father. Now, just as what promised to be a particularly nasty campaign began to heat up, Hillary thought it was time to begin Chelsea's political indoctrination.

Over dinner one evening, Hillary announced that if Daddy lost the coming election, the family would have to pick up and leave. If Daddy won, they could continue living in the only home Chelsea had ever known. Chelsea had to be prepared for the fact that Daddy had "enemies" and that they would say "terrible things" about him. "They might even lie," Hillary told her daughter, "just so people will vote for them instead of Daddy."

Hillary suggested they play a game in which Chelsea played her father on the campaign stump. "My name is Bill Clinton," she said proudly. "I've done a good job and I've helped a lot of people. Please vote for me."

Chelsea was unprepared for what happened next. While the little girl waited for her parents to tell her she had done a wonderful job stating her father's case, Daddy glared at her. "Well, Bill Clinton, I think you've done a lousy job," he barked. "You've raised taxes and you haven't helped people at all. Why, you are a very mean man, and I am NOT voting for the likes of *you*."

Chelsea burst into tears, but Hillary's "role-playing" continued. Over the next few weeks, both parents fired off questions and pelted her with insults until she was inured to anything negative that might be said about Mommy and Daddy.

These dinnertime drills "helped Chelsea to experience, in the privacy of our own home, the feelings of any person who sees someone she loves being personally attacked," Hillary explained. What was important, she went on, is that Chelsea achieved a "mastery over her emotions" that—in theory, at least—made her impervious to attack.

For years, Hillary spoke of these grueling indoctrination sessions with pride—until she realized that some people regarded them as a form of psychological child abuse. In her memoirs, Hillary would dispense with the subject in two brief sentences.

Hillary also drummed it into Chelsea's head that Republicans—Ronald Reagan and George H. W. Bush in particular—were mostly "rich people who just don't care" about the problems of average Americans. On a trip to Washington during the Reagan administration, Chelsea asked her mother if they could take a tour of the White House. Absolutely not, Hillary replied. "We'll have to wait until someone decent lives there."

As a practical matter, both Hillary and Bill were far too busy to spend very much time with their daughter. In 1987 Hillary, more determined than ever to maintain the alliances she had forged over the years with a wide range of "progressive" groups, took over as chairman of the Manhattan-based New World Foundation. During the two years Hillary headed up New World, the number of left-wing organizations supported by the foundation jumped dramatically. Hillary personally signed off on generous grants to such sterling organizations as the Committee in Solidarity with the People of El Salvador (CISPES), which financed El Salvador's Communist guerrillas, and the Christic Institute, a radical fringe group that advocated "legal terrorism" against retired military and intelligence officials. Perhaps most alarming was the New World Foundation's contribution of $15,000 to Grassroots International, which then channeled the money to two groups affiliated with an organization that Hillary still apparently viewed favorably: the PLO.

In an attempt to appear more centrist, Hillary would later downplay her involvement at the New World Foundation. But back in 1988, she boasted in the group's 1988 report that New World had "turned increasingly to the support and development of progressive activist organizations." Moreover, she described her strategy of making "mostly general support grants, rather than project grants, so as to provide core support for organizers and advocates." In other words, more freedom for the El Salvador guerrillas and PLO affiliates to spend the money however they saw fit.

Notwithstanding her involvement in a wide range of organizations across the country, her work at Rose Law, and her responsibilities as Arkansas's First Lady, Hillary's main focus was still on Part One of The Plan. Opportunity knocked in May of 1987, when Democratic front-runner Gary Hart withdrew from the presidential race after the *Miami Herald* ran photographs of the married senator with Donna Rice in his lap aboard the prophetically christened yacht *Monkey Business.* Ironically, with Hart no longer in the race, many eyes in the Democratic Party turned toward Bill.

Clinton's first visit to New Hampshire as a potential candidate was a huge success, and reports from other primary states were equally encouraging. Hillary would later claim that she had opposed Bill's entry into the 1988 race on several grounds: Republican vice president George H. W. Bush, heir to the Reagan legacy, would be virtually impossible to beat; Hillary's father had suffered a stroke, and he and Dorothy Rodham had moved to a condo in Little Rock so the Clintons could help take care of them; Bill was too young, untested.

In her memoirs, Hillary went so far as to say that Bill was on the fence until Chelsea asked him about their upcoming vacation plans. When her father said he might be too busy running for President to take a vacation, Chelsea replied, "Then Mom and I will go without you." That, Hillary would declare in her revisionist autobiography, "sealed the decision for Bill."

In truth, Hillary pressured her husband to run for President in 1988, and was frustrated over his reluctance to throw his hat in the ring. She had purchased the condo for her parents so that Chelsea's grandparents could care for her while Mom and Dad were on the road.

According to Dick Morris and others, the events that literally overnight brought an end to Gary Hart's career filled Bill with dread. Without letting on to Hillary, he began to quiz friends, advisers, even lovers on the subject of extramarital sex and the impact it might have on a candidate.

Nearly everyone Bill talked to was aware that he had cheated on Hillary—he went so far as to admit it to several of them. But only a handful knew the magnitude of Bill's faithlessness. One who had an inkling was the governor's chief of staff, Betsey Wright.

Like Hillary, Wright had a reputation for being abrasive; she routinely yelled at and swore at underlings because "sometimes it's the only way to get people's attention." Most important, she was someone Bill could count on to give him her unvarnished opinion. A few days before the July 14 deadline he had set for himself to make an announcement, Wright confronted her boss with a list of his rumored lovers. There were at least twelve women on the list, and for the next four hours Wright grilled Bill about his relationship with each one of them. She did not even bother to bring up the subject of his many one-night stands, the nagging question of Danny Williams's paternity, and his alleged dalliances with the prostitutes on Spring Street. Wright's inevitable and sobering conclusion: Bill could not run in 1988.

When Clinton told his wife that Wright had talked him out of running, Hillary was livid. She demanded to know what Wright could possibly have said to make Bill change his mind. Wright hemmed and hawed, unwilling to hurt her old friend by going into specifics. It was enough that Bill admitted to her that he had committed adultery with not one woman à la Gary Hart, but with several.

"These women are all trash," was Hillary's startling reply. "Nobody is going to believe them." She *would see to it* that nobody believed them. The problem was manageable. There was no need to panic like Hart.

Bill knew better. Now, so soon after the Hart scandal, every candidate's sexual life would be under the microscope. In a run against a candidate as strong as Vice President Bush, it was the kind of scrutiny Bill simply could not survive.

Hillary stood next to her husband and wept as he announced to a roomful of stunned reporters that he had decided not to run for President. He waxed poetic about Chelsea, and how he owed it as a father to be there for her. Hillary, her eye already on 1992, had no intention of letting on that their grand plan had been momentarily sidetracked by her husband's sleazy sexual escapades. As far as the public was concerned, Hillary wanted her husband to make this noble sacrifice for the sake of his young family.

She was, in reality, seriously considering divorce. As she faced turning forty, Hillary felt betrayed in a way that she never had before. It was not so much the women, but the fact that her husband's unbridled libido had forced a change in their schedule. They would have to wait another four years before reaching for the ultimate prize. "Hillary had been co-governor for years," said a longtime Arkansas supporter. "She couldn't wait to be co-President."

There was another, more personal issue that now rankled Hillary. As angry as she had been at Betsey Wright for talking her husband out of running, the growing list of Bill's partners was a wake-up call for Hillary. In light of Bill's long-standing aversion to wearing a condom, Hillary now worried that he was putting her health at risk in this age of AIDS and other sexually transmitted diseases. She demanded that he be tested for HIV.

Dorothy Rodham, Diane Blair, and others close to Hillary worried that Hillary was indeed on the verge of ending the marriage. In the end, Hillary, keenly aware of the psychic pain suffered by children of divorce, opted to stay. The shouting matches persisted, however, and Dorothy Rodham worried about what impact the Clintons' pitched battles were having on Chelsea. According to a Rodham family friend, Chelsea, after listening to her mother smash things and scream "bad words" at Dad, sometimes cried herself to sleep.

On those many nights when Dad was nowhere to be seen, Chelsea often heard Mom chatting with her coworker Vince Foster. At times it seemed as if Hillary was spending more time with Foster, a childhood friend of Bill's, than she did with her own husband. A secretary at Rose Law said that as far back as the 1970s, Hillary and Foster behaved "like two people in love"—an observation made by many who knew them.

L. D. Brown was in a unique position to observe what transpired between Hillary and Foster, whom she kiddingly called "Vincenzo Fosterini" because she thought he looked like a suave Mafia consigliere. A particular favorite of Bill's, Brown was engaged at the time to Chelsea's nanny Becky McCoy, daughter of the mansion's administrator Ann McCoy. Bill took an almost fatherly interest in the young trooper, and would later be instrumental in getting Brown a job with the CIA.

According to Brown and other members of the Clintons' security detail, no sooner would Bill leave the mansion than Foster would show up to spend time with Hillary. Often, he would not leave until the following morning.

In a sentiment echoed by many others, Brown was "just amazed at how public they were about their affair. All their friends knew exactly what was going on." On one occasion, Vince and his wife, Lisa, were leaving a restaurant with the Clintons and another couple. While the others walked ahead, Hillary and Foster lagged behind. Brown, walking directly behind them, saw Foster groping Hillary's behind. "He'd be kissing her—and I mean real heavy, open-mouthed, tongue-down-the-throat stuff—and then he winked at me. They didn't care who knew."

At another Little Rock restaurant—this time to celebrate Hillary's birthday—Hillary was seated at the bar when, according to Larry Patterson, Foster grabbed Hillary's behind with both hands and squeezed. Then he gave Patterson a wink and made the "okay" sign with his thumb and forefinger. Moments later, Foster placed his hand over one of Hillary's breasts and again winked at Patterson.

Brown, who was exempt from much of the verbal abuse Hillary heaped on the troopers because of his relationship with Chelsea's nanny, was convinced that Foster and Hillary were not just lovers. "Hillary and Vince were two people who were obviously *deeply* in love," he said. "I saw them locked in each other's arms, necking, nuzzling. . . . Vince was a great, great guy, and he was just totally devoted to Hillary in a way that Bill never was."

According to numerous sources, the affair intensified following Bill's decision not to run in 1988. Now the troopers were driving Hillary and Vince to the mountain resort of Huber Springs, where their law firm owned a cabin. There the couple spent hours alone while the troopers waited outside. "I guess Hillary figured that if we did this for her husband," Brown said, "then we damn well better keep our mouths shut and do it for her."

Revenge played a role in Hillary's relationship with Foster, certainly. "Here was Bill screwing all these women right under her nose," said a former employee of Rose Law who worked with Hillary, "and what was she supposed to do, nothing? Vince and Hillary had a very warm friendship and intellectual respect for each other to start with. There was always affection there, but it probably wouldn't have gone to the next level if it weren't for all the crap she had to put up with from her husband."

There was considerable speculation concerning Lisa Foster, mother of Vince's three children. "Lisa loved Vince," said a neighbor, "but there had to have been some jealousy there. Vince and Hillary were around each other at the law firm, and on top of that he ran to the Governor's Mansion whenever she snapped her fingers. Would any wife be happy having to put up with that?"

In an unguarded moment, Hillary apparently felt compelled to explain her relationship with Foster to Brown. "There are some things you have to get outside your marriage that you can't get in it," she told him. "L.D., sometimes you just have to make a leap of faith." (Years later, after Vince Foster's death, Barbara Walters asked Hillary point-blank, "Were you lovers?" Hillary hesitated. "I miss him very much," she answered, never denying the assertion. "And I just wish he could be left in peace, because he was a wonderful man to everyone who knew him.")

Hillary never allowed her own extramarital activities to cloud her political judgment, however. Hillary had seen to it that Bill threw his weight behind the candidate who would go on to clinch the Democratic nomination, Massachusetts governor Michael Dukakis. Now it looked as if the Clintons would be going to the 1988 convention in Atlanta after all, to deliver the prime-time nominating speech.

As Hillary and Bill stepped up to the podium to essentially make their debut before a national audience, she wondered aloud if the houselights would go down as they had for previous speakers. They wouldn't. She asked if Dukakis's floor leaders would ask delegates to listen. They didn't. And she worried that the speech, expanded and approved by the Dukakis camp, was too long. It was.

For thirty-three unbearable minutes, Bill prattled on as the crowd booed, hissed, and hollered for him to get off the stage. Hillary smelled conspiracy—instantly. Was this, she wondered, a preemptive strike against a Democrat who might give Dukakis competition in the future? "It was one of the most agonizing moments in my life," she said, blissfully unaware of all that was to come, "because I knew we had been misled, and I couldn't figure out why." There was thunderous applause for only one line in Bill's speech: "And now, in conclusion . . ." Backstage after the speech, Hillary watched in wide-eyed silence as her husband accused "that little Greek motherfucker" Dukakis of sabotaging him.

Crushed, Betsey Wright flew home to Arkansas, certain that she had just witnessed the annihilation of her boss's career. Hillary, in a state of shock, returned alone to the Clintons' hotel. Bill lingered on and, according to Larry Patterson, groped an attractive young woman before slipping into a room with her and locking the door. "Sex was his drug," he said. "Hillary had to have known that."

Searching for a way to undo the damage, Hillary enlisted the help of their TV producer friends Harry Thomason and Linda Bloodworth Thomason. Eight days after the convention debacle that had made him the laughingstock of the country, Bill went on Johnny Carson to make fun of himself and play the saxophone. "Yet another comeback," declared Hillary, beaming.

Hillary grew impatient as the 1990 governor's race approached and Bill, bored and edgy after a decade in office, refused to commit to running for a sixth term. Hillary and Dick Morris thought he should. It would be easier to run for President as a sitting governor in 1992 than as a private citizen, they argued, and if he lost he could always return to the Governor's Mansion and wait for the next opportunity.

Betsey Wright, however, quarreled bitterly with Hillary over this issue; she believed that Bill's sex life was a time bomb waiting to explode, and that as governor he would only invite scrutiny by the press. Wright's brusque style had alienated many key figures in Clinton's administration, but it apparently took this falling-out with Hillary to send her packing.

Hillary might not have agreed with Wright's doom-and-gloom prognosis, but she suffered nonetheless. Things came to a head when her husband met and fell in love with a stunning blond divorcée named Marilyn Jo Jenkins. An executive with the Arkansas utilities company Entergy, Jenkins was connected by marriage to one of Arkansas's wealthiest families. Bill and Marilyn Jo would later deny that theirs was anything more than a close friendship, though there was much evidence to the contrary.

When he wasn't visiting her at her apartment at the Shadow Oaks Condominium or spending hours talking to her on the phone, Bill was sneaking Jenkins into the mansion—even as

Hillary slept upstairs. One of the troopers, Danny Ferguson, would later swear under oath that Clinton confessed his love for Jenkins. "It's tough," he told Ferguson, "to be in love with both your wife and another woman."

Betsey Wright viewed Clinton's relationship with Jenkins as "profoundly disruptive." Hillary certainly felt so, and she confronted her husband about this new threat to their marriage. In a dispassionate call to Dick Morris, Bill conceded that his marriage was faltering and wanted to know what impact divorce would have on his political future. "He didn't sound particularly upset, not at all," Morris remembered. "Just very cool, very analytical."

Bill's reluctance to run in 1990 and the emergence of this new threat in the form of Marilyn Jo Jenkins pushed Hillary to the edge of despair. In a last-ditch attempt to save their marriage—and force Bill to deal with what he would later concede was his sexual addiction—Hillary convinced him to go with her for counseling from a Methodist minister. Bill was reluctant at first, until he overheard a comment Chelsea made to her mother while wrapping Christmas presents. "Mommy," she asked, "why doesn't Daddy love you anymore?"

For Hillary, the entire period between 1983 and 1989 had apparently been so painful that she either blotted it out of her memory or consciously chose to ignore it. Out of the 562 pages that make up her autobiography, fewer than five pages are devoted to these six crucial years.

The Methodist marriage counselor apparently worked wonders; Hillary emerged after several emotional sessions to proclaim that she and Bill had recommitted themselves to the marriage. She had no inkling that almost immediately Bill would be back on the phone to Jenkins, and over the course of just three months sneak her into the basement of the Governor's Mansion no fewer than four times.

Still, Hillary managed to convince herself that they were back on track—politically as well as maritally. She also saw a clear road ahead in the quest for the presidency. But first, she would have to deal with her husband's reluctance to run for a sixth term as governor.

Before she was banished from the Clintons' inner circle, Wright had actually floated the idea that Hillary would make a superb governor. Hillary would later claim she had no interest in pursuing the governorship, or any other elective office, for that matter. "I don't want to run for office," she told a reporter. "I had dozens of people call me and tell me to run. But it just wasn't anything I was interested in."

In reality, Hillary was eager to run, and told her husband and his staff she intended to do so if he didn't. Dick Morris was asked to do a poll, and the results were unmistakable: a vast majority of voters did not approve of Hillary running. Disbelieving, Hillary insisted that a follow-up poll be done. Again, voters nixed the idea.

Even when Bill did grudgingly agree to run yet again, it was Hillary who went out to do battle with his foe in the primaries, Rockefeller Foundation policy analyst Tom McRae. As television cameras recorded the encounter, Hillary crashed a McRae press conference and then proceeded to upbraid him for unfairly attacking her husband. The supposedly spontaneous debate—Hillary insisted she was just passing by—had not only been carefully planned but rehearsed by the governor's wife.

Hillary's bold move did little to help her husband, and for a time it looked like he might not even win his party's nomination. Once again, Hillary watched as Bill lashed out at his subordinates and advisers for failing him. In May of 1990 he accused Dick Morris of talking him into running and then abandoning him. "You're screwing me!" he bellowed at Morris, who replied with "Go fuck yourself" before storming out of the room. Hillary, realizing how valuable Morris could be in a presidential race, pleaded with Bill to calm down. Instead, he lunged at Morris and pulled him to the floor. Then, Morris recalled of the incident, Clinton "cocked his fist back to punch me."

At that point Hillary, shouting for Bill to get control of himself, pulled her husband off Morris. As Morris got up and tried to make his escape, Hillary chased after him. "He didn't mean it," she pleaded. "He's very sorry. He's overtired. . . . He hasn't slept well in days." Then she put her arm around the shaken adviser and together they walked the grounds of the Governor's Mansion. "He only does that," she told Morris, "to people he loves." (Two years later, at the height of Clinton's presidential campaign, it looked as if the story of Clinton's assault on Morris might find its way into print. Hillary instructed Morris to "say it never happened.")

Hillary made certain Morris was kept in the loop throughout the 1990 election. But when it was time to consider a run for the presidency, Morris recommended to Hillary that they hire James Carville to manage the campaign.

To get reelected, Bill followed Hillary's advice and pledged not to run for President in 1992. Bill would, the Clintons both vowed to the people of Arkansas, serve out his four-year term as governor. Yet even before her husband trounced his Republican opponent, Little Rock lawyer Sheffield Nelson, Hillary was mapping out a strategy to win the nomination. "She always thought," Bill later said of his wife's early determination, "that the right kind of Democrat would have an opportunity to be elected in 1992—always."

Not even George H. W. Bush's 90 percent approval ratings following the Gulf War were enough to dampen Hillary's enthusiasm. "It was amazing," Bill recalled. "That's one where her instinct was right and I didn't feel that way for the longest time."

Hillary was rightly convinced that President Bush was vulnerable on domestic issues in general and health care in particular. While they were seated next to each other at a dinner, Bush had expressed shock when Hillary informed him that the U.S. placed nineteenth in terms of infant mortality, behind such countries as France and Japan. The next day, he passed a note to Bill. "Tell Hillary," it read, "she was right."

"HRC"—or "Herc," as she was sometimes referred to—quickly assembled a campaign staff made up largely of friends and classmates from Wellesley and Yale. "It was her team, all right," said one volunteer, "right down to that horrible Susan Thomases." A partner in the prestigious New York law firm of Wilkie, Farr & Gallagher and Hillary's most influential staff member, Thomases was described by others inside and outside the Clinton team as tactless, profane, and demanding. A *Washington Post* reporter covering the primaries walked into Clinton headquarters to find a campaign worker holding a phone receiver aloft so Thomases's shrieked obscenities could be heard clearly by everyone in the room.

Hillary also called on Maggie Williams, who had worked with her on the Children's Defense Fund, and Brooke Shearer, the wife of Bill's Oxford roommate (and future deputy secretary of state) Strobe Talbott. These women would form the core of what would be known by friend and foe alike as "Hillaryland."

In the company of these well-educated, like-minded liberal women, Hillary was pleasant, even laid-back. This would be in stark contrast to the way she continued to treat the unfortunate men whose task it was to protect her. On Labor Day morning 1991, Hillary was pulling away from the Governor's Mansion when she noticed that the troopers had not yet raised the American flag. She turned the car around, jumped out, and yelled, "Where is the goddamn fucking flag? I want the goddamn fucking flag raised every fucking morning at fucking sunrise!"

By then, of course, the pressure was getting to her. To head off the inevitable charges of infidelity, Hillary had accompanied Bill to the Sperling Breakfast, a regular gathering of influential Washington journalists. Together, they admitted their marriage had not been perfect, but insisted that they intended to be together "thirty or forty years from now."

What no one in the room—including Hillary—realized was that, just a few months earlier at Little Rock's Excelsior Hotel, Bill had unzipped his pants and demanded oral sex from a twenty-four-year-old secretary named Paula Jones. There had been so many Paula Joneses in Bill Clinton's life, he did not suspect that this one would have far-reaching consequences for his marriage and his presidency.

Hillary also worried that Bill would be slammed for breaking his promise to the citizens of Arkansas that he would serve out his term as governor. Again at Hillary's urging, he barnstormed the state for a week, apologizing to voters for breaking his pledge but explaining to them that as President he could better the nation—and bring federal projects and the jobs that went with them back home to Arkansas.

Scarcely a month after Hillary watched her husband announce his candidacy for President, their campaign was hit with its first sex allegations—and from a source that no one in the Clinton camp anticipated. In a *Penthouse* article entitled "Confessions of a Rock 'n' Roll Groupie," Little Rock's Connie Hamzy described her sexual encounters with the likes of Don Henley, Keith Moon, and Huey Lewis. She also revealed that in 1984 Bill Clinton was giving a speech at a Little Rock hotel when he spotted her lounging by the pool. Hamzy claimed Clinton made a pass, saying "I want to get with you," and some mutual "fondling" ensued. To prove her version of events, Hamzy later submitted to a polygraph test—and passed.

Amazingly, prior to the magazine's publication, Hamzy tipped off Hillary through a mutual acquaintance at the Rose Law Firm. This gave Hillary time to grill her husband on what transpired. Bill made light of the incident, claiming that it was Hamzy who approached him, pulling down her bikini top to reveal her breasts.

Hillary was not amused. "We have to destroy her story," she said bluntly, asking if there was anyone who would debunk Hamzy's version of events. Within hours, Hillary had lined up a state legislator who was willing to go on record as saying that he was there, and that Hamzy was lying.

Hamzy was furious at the campaign to discredit her, and promised to deliver the full story in her memoirs. Conveniently for the Clintons, Hamzy's autobiography never made it to bookstores after an unidentified group of Little Rock "investors" hastily acquired the rights.

With their first official "bimbo eruption"—a memorable phrase coined by Betsey Wright—behind them, the Clintons hit the campaign trail together. From the start, they sold themselves as a team. "If you vote for my husband, you get me," she told crowds. "It's a two-for-one blue plate special."

Not everyone was charmed by the idea of Hillary selling herself as "co-President." Richard Nixon, whom Hillary still viewed as "evil," told the *New York Times* that "if the wife comes on as too strong and too intelligent, it makes the husband look like a wimp." Hillary, confident that Americans loved the idea of a husband and wife sharing power in the White House, dismissed Nixon's cautionary observation. "Tricky Dick" was a bitter old man, she told campaign staffers, who was merely trying to take revenge on one of the people who had helped bring down his presidency.

Nixon was to be the least of her worries. While Hillary was campaigning in Atlanta, she received a call from Bill telling her that "a woman named Gennifer Flowers" was saying they had had a twelve-year-long affair. The story was out in that day's edition of the *Star.* She asked him if the story was true. He said no.

Just a week earlier, Gennifer had acquiesced to Bill's wishes and denied the rumors. Now, Bill told his wife, it appeared Flowers had taken money to change her story. "Then we have nothing to worry about," Hillary said. "No one will believe her."

Trouble was, Flowers claimed she had audiotapes of their phone conversations to substantiate her story. At campaign headquarters in Little Rock, James Carville and Deputy Campaign Director George Stephanopoulos were desperate, and would have been more so had they known that Bill had actually phoned Flowers after the first tabloid story hit the stands.

Hillary's husband seemed oblivious to the fact that his campaign staff was going into a free fall. That same day, she arranged for a conference call from Atlanta and exhorted the troops to "get back to work."

But there was no way to sweep the Gennifer Flowers story under the carpet. Soon Hillary, who years later was still dismissing her husband's affair with Flowers as "a whale of a tale," reached the conclusion that something had to be done before the New Hampshire primary.

Hillary huddled with the Clinton team—including Harry Thomason, political strategist Paul Begala, Stephanopoulos, and Carville—and it was agreed that they appear on CBS's *60 Minutes* to deal with the Gennifer Flowers story head-on. The interview would be conducted that Sunday morning at Boston's Ritz-Carlton Hotel, and air that evening right after the Super Bowl.

The interview with CBS's Steve Kroft had barely begun when a heavy light pole began to collapse. Bill pulled Hillary out of harm's way just as the pole smashed to the floor. Hillary, shaken but also keenly aware that this dramatic moment was being captured on tape, clung to her husband as he lovingly reassured her that she was okay. "I've got you," he whispered to her. "Don't worry. You're okay, I love you."

"It was a tense moment," remembered a CBS crew member. "But was Hillary milking it? Absolutely." Later, she complained when the network decided not to air footage of the accident. "I wish they had," she told her friend Diane Blair. "It would have shown everyone how much we really do love each other."

What CBS did air was Clinton's masterful parsing of words, a technique that allowed him to appear to deny the allegations without actually denying them.

Hillary had agreed with their advisers that the best way to handle the situation was simply to admit that there had been problems in their marriage, and that they had overcome them. But as Kroft kept gently probing for specifics about Flowers, the candidate's wife grew noticeably more impatient.

Kroft asked if Flowers was a friend, an acquaintance. "Does your wife know her?"

"Oh, sure," Hillary interjected.

"Yes," Bill said, looking sheepish. "She was an acquaintance, I would say a friendly acquaintance. . . ."

Later, Hillary joined in. "When this woman first got caught up in these charges, I felt as I've felt about all of these women [*all of these women?*]: that they . . . had just been minding their own business and they got hit by a meteor. . . . I feel terrible about what was happening to them. Bill talked to this woman every time she called, distraught, saying her life was going to be ruined, and . . . he'd get off the phone and tell me that she said sort of wacky things, which we thought were attributable to the fact that she was terrified."

Hillary, who had delivered a thinly veiled threat to Juanita Broaddrick and repeatedly vowed to destroy the credibility of any woman who came forward, managed to make Flowers seem sympathetic if slightly deranged.

Kroft again asked if Bill was categorically denying an affair with Flowers. "I've said that before, and so has she," Bill replied. In other words, previously they had both categorically denied the affair—not that he was actually denying it now.

"The problem with the answer is it's not a denial," Kroft continued.

Bill continued to squirm, and was fumbling for an answer when Hillary interrupted again. "There isn't a person watching this who would feel comfortable sitting here on this couch detailing everything that ever went on in their life or their marriage. And I think it's real dangerous in this country if we don't have some zone of privacy for everyone. . . ."

Hillary did not speak up again until Kroft said that he thought it was "admirable" that the Clintons had come to some sort of "arrangement" in their marriage. Protesting the use of the word *arrangement,* Hillary said, "You know, I'm not sitting here—some little woman standing by my man like Tammy Wynette. I'm sitting here because I love him, and I honor what we've been through together. And, you know, if that's not enough for people, then heck, don't vote for him."

Hillary's "Stand by Your Man" reference was a misstep; Wynette wrote an angry letter to Hillary, and Hillary was forced to publicly apologize to the Nashville legend for offending her and countless country-western fans. Otherwise, Hillary's performance was convincing enough to save her husband's candidacy.

Steve Kroft, for one, had no doubt who was guiding the direction of the crucial *60 Minutes* interview. Hillary, he recalled, "was in control. Hillary is tougher and more disciplined than Bill is. And she's more analytical."

A sense of euphoria swept over the Clinton camp, but it didn't last long. The next day, Hillary was sitting in her hotel suite in Pierre, South Dakota, watching Gennifer Flowers play tapes of her conversations with Bill at a televised press conference. The voice was unmistakably Bill's, and the conversations were hushed, intimate.

Hillary—calm, collected, *analytical*—called Bill and talked strategy. Flowers had reportedly been paid $100,000 to tell her story to the *Star,* and Bill was counting on that fact alone to discredit her. "Hillary, who's going to believe this woman?" he asked.

Of course, none of this could disguise the fact that it was Bill's voice on the tapes—that had been authenticated by an independent laboratory—and that, in addition to the many whispered confidences between them, Bill affectionately called Flowers "baby."

The fact of Bill's infidelity, not to mention his shockingly inept handling of Flowers, were of no concern to Hillary. She arranged for another one of her famous conference calls and roused the troops to action. It was too bad she couldn't get Flowers on the witness stand, Hillary complained. "I'd crucify her," she said.

While Bill flew off to Little Rock to escort eleven-year-old Chelsea to a father-daughter dance, Hillary led the attack on ABC's *Prime Time Live,* where she ignored all the evidence and lamely insinuated that the tapes had been faked. Dolly Kyle Browning, meanwhile, was being warned that she would be "destroyed" if she accepted a similar six-figure offer from the tabloids.

After their brief falling-out, Betsey Wright had been brought back to ride herd on the scandals that threatened to overtake the campaign. Chief among her responsibilities was to nip any more "bimbo eruptions" in the bud. Several private investigators were put on retainer to help out, most notably the San Francisco–based husband-wife team of Jack Palladino and Sandra Sutherland.

First on Palladino's list was former Miss Arkansas Sally Perdue. He contacted dozens of friends and relatives until he unearthed one eager to go on the record attacking her virtue. Palladino was not Perdue's only problem. She was also receiving threatening letters, including one that read, "Marilyn Monroe got snuffed." When the rear windshield of her Jeep was shattered by a shotgun blast, Perdue fled the country.

The skillful use of private detectives, opposition researchers, and other covert operatives would become part and parcel of Hillary's modus operandi. Eventually, Dick Morris observed, Hillary built up a "secret police" for the purposes of conducting a "systematic campaign to intimidate, frighten, threaten, discredit, and punish innocent Americans whose only misdeed is their desire to tell the truth."

Hillary was not surprised that Bill's girlfriends would constitute a major—if not insurmountable—problem. She was blindsided, however, by the release of Bill's 1969 letter to Colonel Holmes thanking him for "saving me from the draft." Clinton denied the *Wall Street Journal*'s allegation that he had actually received an induction notice before pulling strings to get into the ROTC. Hillary believed him. So did James Carville and George Stephanopoulos—until a copy of the induction notice was unearthed. Although Stephanopoulos would later say he was so upset by the notice that he went to bed and hid under the covers, Hillary shrugged it off.

Although Hillary and Bill seldom traveled together during the campaign, she kept a watchful eye on the company he kept. The attractive flight attendants aboard *Longhorn One,* the chartered Boeing 727 that served as the Clintons' campaign plane, had been instructed to wait on the plane while the candidate disembarked. "It is important," she said, "that Bill not be photographed leaving the plane with a stewardess—or any woman, for that matter." To his chagrin, Hillary also ordered the staff to make separate sleeping arrangements for the flight crew. "I don't want these women staying at the same hotel," she said.

That did not stop Bill from continually making passes at the flight attendants—fondling one, Cristy Zercher, while Hillary snored loudly just a couple of yards away. When the stories about what went on aboard *Longhorn One* began to percolate years later, Zercher was contacted by the White House and asked not to talk. Not long after, someone broke into her apartment and stole just two things: Zercher's diary, and photos she had taken of Bill Clinton aboard the plane.

Bill always seemed to find time for his "hobby," in large part because Hillary was, as Vince Foster said, "doing all the heavy lifting." With good reason. With the Clintons' ironically named "Truth Squad" out to vilify any woman who dared to go public about Bill's sexual escapades, Hillary was now a lightning rod for criticism.

While the candidate remained aloof on the subject of Whitewater—the term used to describe both the failure of Madison Guaranty as well as the land development deal itself—Hillary was left to explain that Whitewater had left them $68,900 in the red. "If we did something improper," she said, "then how come we *lost* money?"

No sooner had the Tammy Wynette flap begun to die down than she came under fire for alleged conflicts of interest stemming from her work at the Rose Law Firm. Chief among these was the $115,000 fee paid to Hillary's firm to represent the Arkansas Public Service Commission in the state's nuclear power dispute with Louisiana. The commissioners, it turned out, had all been appointed by her husband. "You know," she shot back, "I suppose I could have stayed at home and baked cookies and had teas, but what I decided to do was fulfill my profession. . . ."

Backpedaling furiously over the next few weeks, Hillary tried to explain that she was not attacking stay-at-home moms. Still, there was no disguising the fact that Hillary had always surrounded herself with professional women and had frequently voiced contempt for "those women like Barbara Bush."

By the spring of 1992, polls were showing that Hillary—whose stump speeches often made it sound as if she were the one running for office—had become a major liability to the campaign. Two thirds of voters disapproved of a First Lady sharing power with her husband, and 26 percent just flat-out didn't like Hillary. That sentiment was reflected in the hundreds of disapproving letters Hillary was receiving. One went so far as to brand her "The Antichrist."

"I adopted my own mantra," Hillary later said of this baptism by fire. "Take criticism seriously, but not personally." In reality, no one took criticism more personally than Hillary. "By the time this is over," she warned her daughter, "they'll attack me, they'll attack you, they'll attack your cat, they'll attack your goldfish."

In stark contrast to Barbara Bush, who adamantly refused to alter her appearance in any way to cater to public tastes, Hillary enlisted the Thomasons' help to revamp her image. Arriving in Los Angeles to campaign for the California primary, Hillary willingly placed herself in the hands of Cliff Chally, who did the costumes for *Designing Women,* and the hairdresser Christophe Schatteman (better known simply as "Christophe").

Hillary emerged from the first all-day makeover session with a new wardrobe of pastel power suits and a shorter, honey blond coif. Gone forever were the trademark headbands that seemed to peg Hillary as a humorless, drab, power-hungry career woman.

Hillary would undergo many transformations over the years; the changes in her coiffure alone would spawn an immensely popular "Hillary's Hair" Web site. But the new, more feminine and sophisticated-looking Hillary also had a new line. Gone were all references to power sharing. At least for the time being, Hillary's most important role was that of wife and mother. (Tellingly, Hillary was not pleased when the new stationery—blue letters on a cream background—came back with HILLARY CLINTON emblazoned across the top instead of HILLARY RODHAM CLINTON. After absorbing Hillary's wrath, a hapless campaign aide placed a new order reinstating Hillary's maiden name.)

When it came time for her husband to accept his party's nomination at New York's Madison Square Garden, Hillary looked on with the kind of adoring smile practiced to perfection by another powerful woman behind the scenes, Nancy Reagan.

Now, with vice presidential candidate Al Gore's wholesomely appealing wife Tipper at her side, Hillary posed for photographers sipping tea at the Waldorf, and even competed against Barbara Bush in a cookie bake-off. Hillary gushed appropriately when she won, but failed to mention that she didn't actually bake her cookies. A friend's cook did.

Away from the prying cameras and microphones, it was a very different Hillary who acted as the campaign's chief strategist, head cheerleader, and, when necessary, high executioner. Though she would later shy away from taking credit for it, it was Hillary who first called Clinton's campaign nerve center—which took up the entire third floor of Little Rock's old Arkansas Gazette Building—"The War Room." And she meant it. "Hillary is strictly a take-no-prisoners kind of person," said one top aide. "The world to her is divided into two groups: our friends and the people we want dead."

Hillary valued loyalty above all else, and did not shy away from coming down hard on even senior staffers if she felt they weren't working all-out for the cause. Not even Bill was immune. Hillary nagged him mercilessly about his weight ("For God's sake, try and exercise a little self-control, Bill") and scolded him constantly for not demanding enough from his staff. Oblivious to the presence of others in the room, Hillary often stormed up to Bill, shoved her finger in his face, and let fly with a string of paint-peeling epithets.

Yet, as the election drew near, Hillary and Bill portrayed themselves as the very picture of marital bliss. "If there was a camera within a thousand yards," said one campaign worker, "Hillary would squeeze his hand or whisper in his ear. Once I was standing close enough to make out what she was saying. 'Just keep smiling until these assholes get their pictures.' She never stopped smiling. . . ."

The minute they were out of camera range, "Hillary didn't just let go of his hand. She practically threw it back at him. You didn't get the feeling he liked the lovey-dovey stuff any more than she did."

To be sure, there was the occasional tender husband-and-wife moment witnessed by members of the Clintons' inner circle. "I saw them nuzzling a couple of times," said one, "and he even talked baby talk to her. But even then you had to wonder if this was for our benefit, too."

The affectionate embrace seemed genuine enough when, at 10:47 P.M. on Election Day, all three television networks declared Bill Clinton the winner with 43 percent of the popular vote (Bush received 38 percent, and independent candidate Ross Perot 19 percent). Sitting at the kitchen table in the Governor's Mansion—now master control for the President-elect's transition team—Hillary weighed in on every major appointment, starting with the selection of Texas senator Lloyd Bentsen as treasury secretary.

Hillary resigned from the Rose Law Firm and began "grappling," as she put it, with what her role would be in the new administration. Early in the campaign she had talked to Bill about a cabinet post—until it was pointed out that antinepotism laws passed after JFK appointed his brother Bobby attorney general made such an appointment impossible. Hillary then told her husband she wouldn't mind being his chief of staff, the job that, in effect, she had been doing ever since he went into politics. But Dick Morris and others pointed out that this would put Bill in a difficult position, since he had to feel free to fire his chief of staff if necessary. Hillary then asked if he would appoint her chief domestic policy adviser. But Bill's senior staff, reluctant to see her appointed to any formal position, quickly shot that down as well. (No matter. Hillary's longtime aide, Carol Rasco, got the job. Rasco made certain that Hillary was kept very much in the loop.)

Hillary resigned herself to, as she put it, "having a 'position' but not a real job." There were, she hastened to add, "no laws to prevent me from continuing my role as Bill Clinton's unpaid adviser and, in some cases, representative."

And, in other cases, his probation officer. During the staged farewell ceremony at Little Rock Airport, Hillary noticed that Trooper Patterson had ushered one of Bill's girlfriends onto the tarmac. "What the fuck do you think you're doing?" she asked Patterson. "I know who that whore is. . . . Get her out of here!"

Tensions between Hillary and Bill mounted once they arrived in Washington. On the eve of the inauguration, members of the staff at Blair House cringed as the next President and First Lady argued loudly about office space. Hillary had been led to believe that she would be getting the West Wing Office traditionally occupied by the Vice President. Not surprisingly, Al Gore protested, and Bill reneged on his promise to Hillary.

Yet Hillary was not about to settle for dominion over the residential and ceremonial East Wing, traditional First Lady territory. If she couldn't have the Vice President's office, then she insisted on and got one next door to the White House counsel's office—and directly above the Oval Office. The First Lady would also maintain a large suite in the Old Executive Office Building. In previous administrations, portraits of the President and Vice President were displayed side by side. But like Evita, Hillary decreed that *her* portrait—rather than Al Gore's—hang next to the President's.

The strain was especially evident the next morning at Blair House, when for the first time Hillary—not the chronically tardy Bill—was holding up the show. As he waited outside, Bill shook his head and muttered to himself, "That fucking bitch." The comment was not picked up by microphones but clearly audible to those standing nearby. So was Hillary's reply, delivered as she slid into their waiting car, "You stupid motherfucker!"

A few days after the inauguration, Hillary flew to New York to get some advice from another First Lady driven to the brink of divorce by a philandering spouse. Jacqueline Kennedy Onassis had been enamored of the Clintons ever since she saw the footage of a sixteen-year-old Bill shaking JFK's hand in the Rose Garden. Now, over lunch at Jackie's elegant Fifth Avenue apartment overlooking Central Park and the Metropolitan Museum, Hillary took pointers on how to deal with her loss of privacy, and how to go about giving Chelsea something approaching a normal childhood.

"You've got to protect Chelsea at all costs," Jackie said. "Don't let her think she's someone special or entitled. Keep the press away from her if you can, and don't let anyone use her."

For Chelsea, the adjustment to life inside the White House was remarkably smooth—in large part because of the constant presence of her live-in nanny and companion Helen Dickey. The First Daughter invited friends for sleepovers, ordered take-out food from the local Domino's, rode her bike around the White House grounds, and gobbled down popcorn as she watched first-run features in the White House theater.

The public outcry was predictable when Hillary, that self-proclaimed champion of public education, sent Chelsea to Sidwell Friends, a Quaker private school. It was the only way, Hillary claimed, they could protect Chelsea from intrusive journalists and paparazzi.

Taking Jackie's advice to heart, Hillary went further, essentially barring the press from any access to Chelsea. The ploy worked. For the next four years, levelheaded, even-tempered, well-adjusted Chelsea would not only survive life in the White House fishbowl, but flourish.

Chelsea's mother and father would have a much less easy time of it. Just days after the inauguration, respected White House physician Dr. Burton Lee refused to give the President an allergy shot without first seeing his medical records. Hillary ordered Dr. Lee fired on the spot, and gave him two hours to clear out his office.

The abrupt dismissal of Dr. Lee raised questions about Clinton's medical history that would never be fully resolved; during his entire tenure in the White House, he would never open his medical records for scrutiny, as so many of his predecessors had. It also sent a message that the new First Lady demanded nothing less than total loyalty—and blind obedience.

Unlike other First Ladies, who downplayed their influence, Hillary went to considerable lengths to let people know that, with the exception of her husband, she wielded more power than anyone in the executive branch. She insisted that her chief of staff, Maggie Williams, also be named special assistant to the President—in essence the First Lady's personal emissary at all high-level meetings.

Williams, who had done work for the Children's Defense Fund (CDF), and domestic policy adviser Carol Rasco were not the only FOHs (Friends of Hillary) to occupy key posts in the administration. Former CDF Chairman Donna Shalala was tapped by Hillary to be the new secretary of health and human services, while Hillary's Wellesley pal Eleanor Acheson was put in charge of all nominees for the federal bench and FOH Margaret Richardson was named head of the Internal Revenue Service.

Hillary also rewarded her two closest colleagues at the Rose Law Firm: Webb Hubble, given the number three spot at Justice, and Vince Foster, who as deputy White House counsel would occupy the office next to Hillary's. At first Foster, whose roots were deeply planted in Arkansas, declined. But Hillary's pleas were so impassioned ("We need you, Vince. I need you") that he ultimately relented. It was a decision he would almost immediately regret.

More senior White House aides were assigned to Hillary than to Al Gore—and with reason. She sat in on staff meetings, controlled the President's schedule, and interviewed job candidates. The First Lady "has in many cases served functionally the way a Chief of Staff would in terms of accountability and discipline," a top administration official told noted Washington journalist Elizabeth Drew. "She has made the point openly in his presence. What she does privately I can only imagine."

Her own no-nonsense management style notwithstanding, Hillary would veer along with her husband from one fresh scandal to another. Even before the Clintons moved into 1600 Pennsylvania Avenue, they had to contend with the "Nannygate"—the disclosure that Bill's attorney general designate, Zoë Baird, hired illegal aliens for her household staff and then failed to pay their Social Security taxes despite a $500,000 annual income. The Clintons' next candidate for the job, Playboy-bunny-turned-New-York-jurist Kimba Wood, withdrew her name when it was revealed that she had also hired an illegal immigrant. Janet Reno, the dogged chief prosecutor of Dade County, Florida, finally passed muster, fulfilling Hillary's wish that a woman hold the nation's top law enforcement job.

Increasingly, it became clear that Hillary was in fact the source of many of Bill's problems during that first year in office. Bill ran into trouble again when he followed Hillary's advice and lifted the ban on gays in the military. When she summoned her hairdresser to Los Angeles Airport to give Bill a $200 haircut aboard *Air Force One,* the President was criticized for causing an air-traffic tie-up.

Hillary also concocted the plan to fire longtime White House travel staffers and replace them with Catherine Cornelius, a distant relative of Bill's, and an agency owned by their Hollywood pals the Thomasons. She also launched an investigation into allegations—which later proved baseless—that the Travel Office holdovers from the Reagan and Bush administrations had been taking kickbacks.

When someone questioned the wisdom of axing reputable White House staffers for the apparent purpose of diverting government business to their cronies, White House assistant to the President David Watkins warned there would be "hell to pay if we failed to take swift and decisive action in conformity with the first lady's wishes."

"Travelgate" would trigger yet another investigation into the administration's motives and methods. It would also ultimately claim the life of someone Hillary loved dearly.

It was one thing for Hillary to summarily dismiss the entire Travel Office staff, and quite another to impose her will on the Secret Service. Not that she didn't try. During the transition, members of the Secret Service detail assigned to guard the President-elect (code name: "Eagle") were shocked by the Clintons' behavior. Unlike the state troopers who had done his bidding for years, the agents were not willing to participate with Clinton in rating women, much less approach them on his behalf. Nor were they willing to caddy, go on shopping errands, or carry baggage—tasks that might prevent them from concentrating on their principal job: to protect the President. When an agent explained this to Hillary (code name: "Evergreen") after their plane touched down in Arkansas, she looked him over carefully. "If you want to remain on this detail," she said, "get your fucking ass over here and grab those bags."

The agents were equally disturbed by the Clintons' predilection for squabbling in public. William Bell, a former member of the White House detail, recalled yet another pitched battle as Bill and Hillary rode together in a limousine—he in the front, she in the back. Once again, an enraged Hillary flung a briefing book at Bill, inadvertently hitting the agent behind the wheel in the back of the head.

Things only seemed to get worse in the White House, where early on Secret Service agents watched in amazement as Hillary picked up a lamp and hurled it at her husband. Fortunately, she missed. When word of the incident was leaked to the press, Hillary banished not only the Secret Service but the domestic staff from the residential floors when members of the First Family were there.

Although Hillary claimed she was barring staff from the residence to afford Chelsea (Secret Service code name: "Energy") a level of privacy—something that hadn't concerned them during their twelve years in the Arkansas Governor's Mansion—she suspected that many of the agents resented their new bosses. To some extent, she was right. Several viewed Clinton as a draft dodger and antimilitary, a spoiled baby boomer given to childish, self-indulgent tantrums. Hillary was seen as equally foul-tempered, with the kind of longshoreman's vocabulary they found unbecoming in a First Lady.

Then there were the scores of Clinton employees whose criminal records, histories of drug use, and otherwise shady backgrounds would have normally made them security risks and ineligible to work in the White House. Circumventing long-standing screening procedures and FBI background checks, the White House counsel's office gave permanent passes to more than one hundred Clinton appointees who had been denied them by the Secret Service. Fully twenty-one of those asked to take drug tests failed—a number that included top aides to both Bill and Hillary. Instead of being fired on the spot, they were simply asked to undergo testing twice a year—at a time of their choosing.

Hillary had agreed with the counsel's opinion that drug abusers suffered from a disability and thereby had a *right* to employment at the White House. This was in keeping with her role as Hillaryland's resident mother hen. In that capacity, the First Lady regularly popped into her staffers' offices to compliment them on their appearance, quiz them about their families, and share gossip. She also presided over birthday parties and baby showers, and at times took obvious pleasure in playing matchmaker.

"It doesn't sound very liberated," said a junior staffer, "but these were her girls. She genuinely cared about them so long as she was convinced they were totally loyal to her and the President. If you wavered for just a second, you were dead. Not fired. You just ceased to exist in her eyes and in her mind. We started calling it 'Hillary's Alzheimer's.' She'd be looking right at you, but you weren't there as far as she was concerned. Disloyalty—or anything that could remotely be interpreted that way—was the one unforgivable sin."

How Hillary viewed her husband's disloyalty was another matter. It was not as if he had changed his ways. Now that he resided in what he called "the crown jewel of the federal prison system," Bill surrounded himself with attractive women who, it was widely assumed, went beyond their job descriptions in serving the President. One of the flight attendants aboard Clinton's campaign plane, Debra Schiff, landed in the West Wing as a receptionist. A "knockout," as chief White House steward Mike McGrath described her, Schiff favored a working wardrobe of short skirts, tight sweaters, and high heels. Schiff, who would eventually deny having an affair with Clinton, nonetheless spent twenty minutes with him alone in the Oval Office study every morning. Why? "You figure it out," she replied. Secret Service agent Gary Byrne didn't have to; one morning he stumbled upon Schiff and the President enjoying each other's company.

Schiff was by no means alone. The equally blond and stunning Catherine Cornelius, who was only nineteen when she joined the White House Travel Office, not only accompanied the President when he traveled without Hillary but had unusual access to the Oval Office.

White House aide Robin Dickey denied having an affair with Clinton, although members of Bill's Little Rock security detail swore under oath that she admitted to them that she had. Dickey raised eyebrows when she repeatedly showed up at the Oval Office to massage the President's back.

Of all Bill's alleged White House paramours, none was more problematic for Hillary—at least during the Clintons' first term—than Marsha Scott. The daughter of Philadelphia Eagles football star Clyde "Smackover" Scott and a former Miss Arkansas, Marsha liked to refer to herself as Bill's "old hippie girlfriend." Scott, now director of White House correspondence, would eventually admit to spending many nights alone with the President in the family residence.

The occasional blowup aside, for the most part Hillary turned a blind eye to her husband's indiscretions. There were, after all, other, more pressing demands on her time. To sell her plan for universal health coverage linked to cost control and preventive health care, Hillary held public hearings, huddled with legislators, and barnstormed the country denouncing the insurance and pharmaceutical industries.

Hillary, who insisted that much of the process take place behind closed doors and away from scrutiny by the press, quickly came under attack for violating sunshine laws. She also refused to invite input from special interest groups, whose support would be needed if there was any hope of getting her Task Force on National Health Care Reform through Congress. This, coupled with Hillary's blanket refusal to compromise, eventually doomed the Clinton health care package.

More than a humiliating defeat for Hillary, the gradual implosion of her highly touted health care initiative would embolden those who had vocally opposed the Evita-like concept of a husband-wife co-presidency. Now she faced more inquiries about Whitewater and the collapse of Madison Guaranty. Amid talk of a possible indictment, Hillary broke down during a meeting of Bill's inner circle. "I know everybody's looking out for Bill," she said, choking back tears, "but nobody's out there fighting for *me.*"

No one sympathized with Hillary's Whitewater woes more than Vince Foster, who suffered an anxiety attack when Zoë Baird bowed out of contention for the post of attorney general. Foster blamed himself for failing Hillary, and slid deeper into depression as the scandals piled up.

Nevertheless, according to Secret Service agents assigned to protect the First Family at the time, the affair between Hillary and Vince Foster continued at the White House. One claimed to have seen them locked in a passionate embrace inside Foster's office; another would only acknowledge that he had witnessed enough to convince him that "it was still going on between them, definitely." The relationship between Hillary and Vince was the subject of endless gossip and speculation among women at the White House, who voted Foster the staff member they would "most like to have an affair with."

In addition to the plethora of scandals, Hillary was dealt another blow in March of 1993 when Hugh Rodham suffered a massive stroke. While she was at her father's bedside in Little Rock agonizing over whether or not to take him off life support, Bill was entertaining his friend Barbra Streisand at the White House. It was only after her father died two weeks later and she returned to Washington that Hillary learned that Streisand had spent the night in the Lincoln Bedroom.

Hillary was upset that her husband was spending time with one of Hollywood's biggest stars while she kept a vigil at the bedside of her dying father. The day Hillary returned to Washington, a steward who had been summoned to the second floor made a hasty retreat when he heard shouting and the unmistakable sound of slamming doors. At the following morning's news briefing, reporters peppered Press Secretary Dee Dee Myers with questions about the wounds clearly evident on the President's face and neck. Before she ever saw the claw marks, Myers told reporters that Clinton had simply cut himself shaving. "Then I saw him," Myers later admitted, alluding to the fact that the First Lady was clearly disturbed by Streisand's presence alone with her husband in the family quarters. "It was a big scratch, and clearly not a shaving cut."

Apparently, no one had made it clear to the President which story they were going with. Only hours later, Clinton claimed he had injured himself "rolling around" with his daughter on the floor. Hillary, meantime, barred Streisand from staying at the White House.

Over the years, the President would suffer numerous unexplained bruises and abrasions, including a sizable goose egg on his forehead. His explanation: "Walked into a door."

Presidential handlers laughed off any suggestion that their boss's wounds might have been the result of anything other than an innocent mishap. But privately, they also wondered aloud if Hillary, whose penchant for aiming objects at her husband's skull had been well established, was actually capable of spousal abuse.

Even if the First Lady was battering her husband, there were those who felt he well deserved it. Frankly, Betsey Wright observed, "Hillary's tolerance for some of his behavior just amazes me." As dependent as Bill was on Hillary's decisiveness and her organizational skills—the First Lady's influence was "overwhelming," said Clinton's then-adviser David Gergen—the President made little effort to disguise his feelings of resentment toward her. George Stephanopoulos recalled how the mere mention of Hillary's name was like "the flipping of a switch." Aides had no trouble telling when the President was parroting his wife's opinion, which was often. "Even if he was yelling," Stephanopoulos said, "his voice had a flat quality, as if he were a high school debater speeding through a series of memorized facts."

Early in the day, Gergen observed, the President would be upbeat, "even chipper. . . . As we started to work, his mood would darken, his attention would wander and hot words would spew out. WHAT, I would wonder, had she said to him now?"

Bill was not the only one whose mood changed in Hillary's presence. According to several staffers, Hillary often cut off discussions with "I don't recall asking your opinion," or simply, "Who in the hell asked *you?*" Veteran Clinton watcher Joe Klein observed that, if someone questioned a strategy of Hillary's, "the First Lady would, with cold fury, tell the questioners to stuff it—Hillary's plan was THE plan. Bill sat quietly as she bludgeoned respected members of the administration into silence."

Still, at her father's funeral, an emotional Hillary leaned heavily on the two most important men in her life: Bill Clinton and Vince Foster. She would continue to grieve for months, occasionally breaking down in the middle of a staff meeting or a speech.

Just as often, Hillary's grief turned to rage. She felt that her aides—especially the friends and associates she brought along from Little Rock—were letting her down. In much the same way that her tantrum-prone husband vented his frustrations on Stephanopoulos virtually every morning, Hillary now complained bitterly and often to forty-eight-year-old Vince. At a May 13, 1993, meeting she snapped at him, "What's going on? Are you on top of it?"

In his notes from the meeting, Foster attributed Hillary's foul temper to "general frustration" that was "aggravated by grief" over the death of her father. She was apparently less concerned with her friend's well-being, failing to notice that since moving alone to Washington (his wife Lisa remained behind in Little Rock so that their youngest son could finish his junior year in high school), Foster was lonely and suffering from overwork.

Vince was becoming increasingly afraid that he had failed Hillary, and actively feared that he might be called to give testimony that would damage the First Lady—either that, or lie under oath to protect her. He also came to believe that his own reputation was being sullied in the process of defending his old Arkansas friends. A colleague described him as becoming "frantic" when the *Wall Street Journal* ran a piece titled "Who Is Vince Foster?" and illustrated it with a question mark inside a silhouette. Where Hillary lashed out at her attackers, Vince tended to internalize everything. Each article was, as far as he was concerned, a politically motivated attack on his character. "Before we came here," he said of the Clintons' Arkansas friends, "we thought of ourselves as good people."

There were personal pressures as well. The Foster marriage was shaky, mainly because of Vince's intense relationship with Hillary. Lisa Foster had agreed to relocate to Washington if that's what it would take to save her marriage. But by the time she finally joined him in early June, Vince was physically and emotionally wrung out, and on the verge of resigning his White House position.

Comic relief of a sort arrived on Father's Day, when a man named Henry Leon Ritzenthaler told the *Washington Post* that he was the President's long-lost half brother. William Jefferson Blythe, it seemed, had been married at least twice before he met Bill's mother—something no one in the family had known. A few months later, Sharon Pettijohn stepped forward to say she was Clinton's half sister by yet another of Blythe's wives. These paled in comparison to Bill's stepsister Diane Welch, whose father was Virginia's third husband, Jeff Dwire. Welch was a convicted drug dealer and bank robber, and her son, a convicted forger, was a member of the Ku Klux Klan.

With every new revelation about her husband's twisted family tree, Hillary cringed with embarrassment. "Oh, I'm so sorry," Vince Foster commiserated when Hillary told him about Ritzenthaler.

"So am I." Hillary sighed. "You know, I'm just so sick of this."

It was the last time she spoke with Vince. On July 2, two of Vince's coworkers—William Kennedy III and David Watkins—received official reprimands for the part they played in Travelgate. Vince was profoundly embarrassed that, as a result of his special relationship with the First Lady, he was spared. Vince went to his boss, insisting on taking his share of the blame.

Vince's mea culpa fell on deaf ears, as did his tacit pleas for help. The First Lady, fuming over the way Travelgate had been bungled, refused Vince's requests for a meeting and did not return his calls. Foster, who had selflessly done Hillary's bidding for sixteen years, felt betrayed and abandoned. He also worried that he, like so many other Clinton associates, would wind up in jail. Foster began referring to Hillary with no small degree of sarcasm as "The Client."

In early July, Hillary took her widowed mother along as she accompanied Bill to the G-7 summit of leading industrial nations in Tokyo. On the way back, they met up with Chelsea in Hawaii, and then flew on to Arkansas to spend more time with Dorothy Rodham in Little Rock. It was after 9 P.M. on July 20, 1993, when the phone rang at Rodham's condominium. Hillary took the call in her mother's kitchen. White House Chief of Staff Mack McLarty was on the line, and he had terrible news: Vince had killed himself.

When she emerged from the kitchen, Hillary was at first catatonic. Then she began sobbing with such intensity that her press secretary, Lisa Caputo, could imagine only one thing: someone must have assassinated the President. No, but another man Hillary had loved was suddenly dead, and by his own hand.

Even at this traumatic moment, there was a practical question only Hillary could deal with. Bill was, at that very moment, on CNN's *Larry King Live* from the West Wing and had just agreed to stay on for an extra thirty minutes. McLarty wanted to know if he should let the President finish the show before telling him.

"No, no," Hillary answered. "He can't be told while he's on the air. Have him cut the interview short and tell him right now." When McLarty did tell Bill, it appeared for a split second that the President might sink to the floor.

She stayed up all night, talking with friends on the phone and weeping. Even though she must have always known the answer, Hillary kept asking it over and over that night and in the days to come: "Why, Vince? Why?"

He did it.
He really did it.

—To Harry Thomason,
after learning of Vince Foster's death

■

It's my fault.
It's *all* my fault.

I'm not a good actress.
My feelings show on my face.

4

Hillary went over the events of that day again and again, trying to unravel the mystery of Vince's death. Vince had lunch at his desk in the White House, then left around 1 P.M. Then he drove his gray Honda Accord to Fort Marcy Park in McLean, Virginia, got out of the car, sat down on the grass, took out his father's .38 caliber Colt pistol, put the barrel in his mouth, and squeezed the trigger.

She was crushed and confused, but Hillary did not call Lisa Foster to offer words of sympathy. Instead, just forty-five minutes after McLarty's call, she phoned Maggie Williams and dispatched her to Vince's office. If there was anything there that might prove embarrassing to the Clintons—like a suicide note in which he claimed his undying love to Hillary—they had to get their hands on it before the police did. When she opened the door to Vince's office, Williams saw Patsy Thomasson sitting at his desk. A former assistant to convicted coke trafficker Dan Lasater and now White House aide David Watkins's deputy, Thomasson was already on the case.

"Do you know anything about why Vince would do this?" Hillary asked Williams. There was panic in her voice. "What's going on? You know, is there something you can't tell me? What's happening?"

Investigators would soon be demanding answers to the same questions. But the careless handling of evidence at the scene, disappearing (then mysteriously reappearing) files, inconsistencies in the testimony of witnesses as well as stonewalling by the White House made the circumstances surrounding Foster's death all the more suspect.

Conspiracy theories abounded even before associate counsel Steve Neuwirth fished twenty-eight scraps of yellow paper out of Foster's briefcase nearly a week after his death. Neuwirth and his boss, Bernie Nussbaum, hastily put the jigsaw-puzzle pieces together and then asked for Hillary to make sense of it. "I just can't deal with this," she said. "Bernie, you deal with this." Besides, she pointed out, it would not look good for her to appear to be involved.

In the note, Foster insisted that he had never violated any laws related to Travelgate, and that if he committed any mistakes they were largely the result of overwork. But his most damning words were aimed at the press. "I was not meant for the job or the spotlight of public life in Washington," he wrote. "Here ruining people is considered sport."

While Hillary was sobbing with friends over the phone and trying to ferret out whatever information she could, Bill drove to the Foster home to comfort Vince's widow. Then he returned to the White House, where, White House staffer Marsha Scott would tell David Watkins's wife Ileene, Scott spent the night "with Bill in his bed."

At the funeral in Little Rock, Bill comforted Vince's wife and children in the church sanctuary while Hillary was shunted aside. Waiting by herself in an alcove, she took her seat only seconds before the funeral service was to begin. Afterward, the funeral procession drove to Hope, where Vince was buried in a cemetery on the outskirts of town. "By then," she later recalled, "I was beyond words. Numb." Still, Hillary did not linger with the other mourners.

The obvious rift between the First Lady and the Foster family came as a shock to many mourners present. But not to those who knew how deeply Vince and Hillary had felt toward each other, and how hurtful it had all been for Lisa. "She was the stay-at-home mom," said a mutual friend, "and here Hillary was this ultimate career woman, burning up the track. Any wife would be threatened by the kind of closeness Vince and Hillary shared."

Hillary soon learned that Vince had sought psychiatric help and had actually begun taking a prescription antidepressant. This evidence that he had been suffering from clinical depression raised questions about how so many people could have missed the signs. ("Why hadn't anyone noticed? Everyone said he seemed so happy that last day.") But it also offered a way out for Hillary. If Vince's depression was essentially caused by a change in brain chemistry—and not by being suddenly cut off by perhaps the most important person in his life—then Hillary was off the hook. "I will always wish I had read the signs of his despair," she later wrote, "and could have helped him."

Lisa Foster, meanwhile, was angry—especially at the Clintons. Whenever the White House flashed on her television screen, she grew "livid. I hated everything," she later admitted. "I was mad as hell." At the appropriately named Afterthought, a neighborhood hangout favored by the Clinton crowd, Lisa was overheard railing at Bill and Hillary. But within months she was dating widower James Moody, a Little Rock attorney who had also been a friend of the Clintons. After Vince Foster's widow and Moody became engaged in 1995, Bill appointed Moody, whose law firm had contributed heavily to Clinton campaigns, to the federal bench. The couple would marry in 1996, on New Year's Day.

In the aftermath of Vince's suicide, Lisa had somehow managed to come to terms with the continuing rumors of an affair between her husband and Hillary. She had first had to confront them years earlier, at the height of the Gennifer Flowers brouhaha in 1992. Fearing that some enterprising reporter might dig up the truth about him and Hillary, Foster had gathered his family together and warned them that the inevitable stories that would be coming out about him and Hillary were false. "They're going to say we had an affair," Vince said. "I don't see why," he added disingenuously, "I can't be friends with a woman at work without somebody assuming we had an affair."

Lisa accepted her husband's denial. "There are certain things I know, because I was there," she said, "and there are some things I don't know—that I can never prove one way or the other, except by faith. I just have faith in Vince and faith in Hillary that they did not have an affair." It was a subject she never discussed with Hillary. "I mean, would you expect her to deny it?" asked Lisa. "What good does denying do?" Besides, she continued about the rumors of an affair, "If they did, who cares now? You know? Who cares? I sincerely believe that they didn't. But that doesn't matter to me—Vince is dead."

Hillary plunged back into work ("I'm an obsessive personality at heart"), giving speeches on health care and trying to salvage her plan by rallying support in Congress. Still trying to come to terms with the deaths of her father and Vince, she was, in her words, "an absolute basket case." In the middle of one speech, Hillary had to stop to compose herself. Another time, she wiped away tears only seconds before stepping to the podium. But Hillary was just as likely to lash out at those around her. At times, she conceded, she appeared "brittle, and even angry—because I was."

Hillary managed to escape the pressure in mid-August, when she and Bill vacationed on Martha's Vineyard. There they spent time with old friends Vernon and Ann Jordan, William and Rose

Styron, and *Washington Post* publisher Katharine Graham. But the high point for Hillary was the time they spent with Jackie Onassis and her companion Maurice Tempelsman aboard Tempelsman's yacht the *Relemar.* At one point, Bill and Ted Kennedy dared Hillary to follow Chelsea and Jackie's daughter, Caroline, in jumping from the boat's highest platform. It took Jackie to persuade her that she didn't *always* have to do what her husband said.

The vacation in Martha's Vineyard restored Hillary's spirits, and she returned to Washington with a renewed sense of purpose. Things seemed finally to be looking up that fall, as unemployment rates fell along with interest rates and inflation. Housing starts were up, and NAFTA, family leave legislation, and the Brady gun control bill had all been enacted by Congress.

Hillary was unaware that for months Bill had been secretly maneuvering to prevent the Arkansas state troopers from telling their story to the press. The President reportedly went so far as to call Trooper Danny Ferguson and suggest to him that high-paying federal jobs were theirs for the asking—if they kept their silence about procuring women for then-Governor Clinton back in Little Rock.

Not that the threat of yet another scandalous revelation about his sexual past was enough to get Bill to change his ways. On November 29, 1993, the Clintons returned to the White House after celebrating Thanksgiving dinner at Camp David. Looking forward to watching her husband sign the Brady Bill into law the next day, the First Lady had no way of knowing that at 3 P.M. that day Bill would be in the Oval Office study, groping the breasts of a part-time White House volunteer named Kathleen Willey.

On December 17, Hillary was hosting a holiday reception at the White House when she received a call from David Kendall, the lawyer hired by the Clintons to handle the Whitewater mess. Kendall warned her that the *Los Angeles Times* and *The American Spectator* were each publishing articles "full of ugly lies" about

Bill—most notably the Arkansas state troopers' allegations that they had procured women for him.

"As far as I can tell," Betsey Wright said after reading the *Spectator* article, "they're telling the truth." For Hillary, who now worried that Chelsea was old enough to understand exactly what was being said about her father, it was just one more *thing* to deal with. "I am just so tired," she said wearily, "of all of this."

Not too tired to lash out at her husband's accusers, however. Her husband's political enemies were behind the "outrageous, terrible" stories, said Hillary. She also cited "financial gain" as a motive—just as she had with Gennifer Flowers—and said it was "pretty sad" when people resorted to such tactics during the holiday season. This "stuff," as she referred to the troopers' well-documented, first-hand, and largely corroborated accounts of Bill's rampant womanizing, "will end up in the garbage can where it deserves to be."

Once again, Hillary directed her rage not at Bill, but at the press for running with the troopers' stories—and at what she would eventually describe as a "vast right-wing conspiracy" to bring the Clintons down. According to friends, Hillary figured that she had already forgiven him for the sins committed in Little Rock; the public disclosure of Bill's pathological skirt chasing did not change that. But she was upset with her husband for being foolish enough to actually call up the troopers hinting at jobs for silence. Done without the knowledge of Stephanopoulos, Carville, or any other senior advisers, it was a particularly sloppy attempt at a cover-up because it could easily be proven using White House phone records. "Shit, Bill," she said at one point, "even Nixon wasn't that stupid."

Troopergate was still raging when the Clintons were awakened in the middle of the night on January 6. Bill's stepfather, Dick Kelley, had terrible news: the President's mother had died in her sleep from breast cancer at her Hot Springs, Arkansas, home. Bill's strong attachment to Virginia ("Ginger" to Chelsea) was not unlike that of Elvis Presley to his mother; it was a parallel Clinton and his

Elvis-obsessed mom often made themselves. Hillary took it upon herself to walk down the hall, wake Chelsea up, and bring her back to her parents' room so they could both break the news to her.

Hillary had been hit hard by the deaths of her father and Vince only months before. Now, with the loss of Virginia, the Clintons were once again united by grief. "It all had a cumulative effect," a Little Rock friend said. "He'd been there for her when her dad died, and now she was comforting him while he cried about his mother. . . ." Hillary would later say that "sharing the loss of our parents" was one of the things that cemented their relationship, and made divorce unthinkable.

Even on the day Virginia Clinton was laid to rest next to Bill's biological father, Bill Blythe, in Hope, there were calls on Capitol Hill for a special prosecutor to investigate Whitewater. Too upset to confront the special prosecutor question, Bill took off on the night of his mother's funeral for a long-planned trip to Europe to discuss the future of NATO. Hillary, left holding the Whitewater bag, argued forcefully against the appointment of a special prosecutor on the grounds that, among other things, any such investigation would likely drag on for years.

Conversely, most of her husband's advisers felt that by requesting the appointment of a special prosecutor, the President would look as if he had nothing to hide. To make this happen, however, the President's staff members knew they had to win Hillary's backing. Over the course of several days, they actively lobbied Hillary to go along with a special prosecutor, or risk bringing down her husband's administration.

Unconvinced, Hillary gathered senior staff members together for a conference call with the President in Prague. As she always did, Hillary took charge and, when the meeting was through, wrapped up with a summation of both sides. Throughout, in language that others at the meeting would charitably call "colorful," Hillary drove home the idea that a special prosecutor would feel compelled to keep going until he found something. In the end, Bill overruled his wife and instructed her to get the ball rolling. "Why don't you sleep on it, Bill, and—"

"No!" he shot back over the speaker. "Let's get this over with."

Hillary felt she had failed Bill, that she had allowed him to be backed into a corner at a time when he was vulnerable. Worse, she felt powerless to do anything to change his mind. Nevertheless, she went ahead with plans to fly with Chelsea to Moscow, where they would join the President's party. After landing at Moscow's Sheremetyevo Airport, Chelsea got into a car with staffers for the ride into town while Hillary climbed into an official limousine with Alice Stover Pickering, wife of U.S. ambassador to Russia Thomas Pickering. The motorcade was still wending its way through the city when Hillary leaned over—and vomited on the limo floor. The First Lady later blamed the unfortunate incident on the delayed effects of a bumpy landing at the airport.

The stress did not let up after they returned to Washington. During one highly emotional meeting, she again blasted the President's senior advisers for not doing enough to protect her interests. Then she broke down as she told them she was "feeling very lonely right now."

It would be only a matter of weeks before another political grenade was lobbed at Hillary. This time, the young secretary Bill had allegedly approached for oral sex at Little Rock's Excelsior Hotel in 1991 was holding a press conference. In one of its Troopergate articles, *The American Spectator* had identified a woman named Paula who supposedly wanted to be then–Governor Clinton's "regular girlfriend." The piece also suggested that Paula had been a willing participant in a sex act with Clinton. Claiming that her friends could easily identify her from the piece, Jones opted not to sue the *Spectator.* Instead, she sought an apology from the President for making improper sexual advances toward her.

The mainstream press largely ignored Jones's accusations, but that was not enough for Hillary. Dismissing her husband's less-than-polished accuser as "trailer trash," she instructed James Carville and the Clinton legal team to do likewise. "Drag a hundred-dollar bill through a trailer park you never know what you'll find," sniped Carville, not at all self-conscious about his Louisiana bayou drawl or the fact that he worked for a man whose nickname was Bubba.

In keeping with her own lifelong commitment to self-deception, Hillary would never acknowledge that Paula Jones's allegations were totally consistent with the other charges of sexual misconduct leveled at her husband. Years after the Monica Lewinsky scandal led to Bill's impeachment and he confessed to being a sex addict, Hillary was still insisting that it was all part of a right-wing plot to destroy the Clintons. "We expected this story," she wrote in 2003, "to die like the other phony scandals."

Jones upped the ante on May 6, 1994, when she sued the President for $700,000. Bob Bennett, the high-profile Washington lawyer hired to defend Clinton, went even farther than Carville. Bennett publicly compared her to a dog, and hired private detectives to dig up whatever dirt they could in Jones's past.

The public was outraged, and Bennett was forced to qualify his remarks. But Hillary was delighted with the lawyer's over-the-top attacks, and exhorted others on the administration team to follow suit. "These women *are* trailer trash," Hillary said. "They *are* out for money. Why not tell it like it is?"

For an avowed feminist whose husband hailed from one of the poorest states in the South, it seemed nothing short of astounding that Hillary did not hesitate to brand her enemies "bimbos," "tramps," "sluts," "trailer trash," "rednecks," "shit-kickers," and "white trash." Nor were her less-than-politically-correct zingers reserved for Caucasian Southerners. Over the years, Hillary reportedly made anti-Semitic, anti-Asian, and anti-Indian remarks.

Observed a friend from Arkansas who was given a job in the

Clinton administration: "This is a huge blind spot of Hillary's. She would never think of using the N-word or making an antigay remark, but she's tone-deaf when it comes to the feelings of these other groups."

His belligerence aside, Bob Bennett did negotiate with the plaintiff's lawyers for a presidential statement to the effect that Paula had done nothing wrong, and that Bill believed her to be a truthful and moral person. The President, well aware of the facts of the case, found this to be a reasonable compromise and was just about to sign when Hillary weighed in. She insisted that any such statement would be tantamount to an admission of guilt; if Paula Jones was truthful, then her version of events was accurate.

Less than a week after turning down a settlement offer from Bob Bennett, Paula Jones was informed that she was the target of an IRS audit. She was not alone. Under the stewardship of Hillary's Yale Law School classmate Margaret Richardson, the IRS audited a number of conservative organizations and publications—all of which led to charges that the audits were politically motivated.

Hillary made no secret of loathing radio talk-show host Chuck Harder, who frequently delved into Bill's sexual escapades on the air. An IRS audit of his nonprofit People's Radio Network dragged on for over six months, until Harder finally bowed out. Later, when Harder arranged to return to the airwaves with a new radio network backed by the United Auto Workers, Hillary personally called UAW president Steven Yokich and suggested that they hire her brother, Hugh Rodham Jr., to host a talk show instead.

Hillary, meantime, was being warned by her more astute advisers that she had alienated much of the Washington press corps with her closed-door meetings on health reform and her general lack of availability. In April of 1994, Hillary sat down in the State Dining Room for a no-holds-barred press conference. For the next hour she was bombarded with questions about Whitewater,

Madison Guaranty, her 1,000 percent killing in the commodities market, and more.

What impressed everyone more than Hillary's less-than-illuminating answers was her cool demeanor and her deft wardrobe choice. In the ongoing attempt to soften her image, Hillary wore a black skirt and pink sweater set. Reporters quickly dubbed this calculated attempt at damage control the "Pretty in Pink" press conference.

Things were not so pretty in the West Wing, where White House staffers were now accustomed to being verbally abused by both the President and the First Lady. There was a major difference between the two in this regard: like his idol JFK, the President routinely blew up at aides. But once he had vented his anger, Bill moved on. Hillary, in contrast, was not the sort to forgive and forget. She held grudges. "Anybody that stood up" to Hillary, recalled White House Press Secretary Dee Dee Myers, "was, you know, smashed down and belittled, very personally." Myers claimed the President did not attack people personally, but "Mrs. Clinton sometimes did . . . not only would she sort of humiliate you in front of your colleagues or whoever happened to be around," Myers added, "Hillary tended to kind of campaign against people behind their back, and that was certainly my experience."

It was also the experience of Abner Mikva, the retired chief judge of the U.S. Court of Appeals for the District of Columbia who joined the White House counsel's office in 1996. He felt that Hillary was largely responsible for the atmosphere of paranoia inside the White House. When Mikva complied with the courts and turned over subpoenaed documents, Hillary showered him with obscenities. The former jurist, unaccustomed to such behavior, resigned.

As time-consuming as the plethora of scandals had become, there were other matters for America's co-Presidents to contend with. In May, the U.S. was presented with an opportunity to arrest one of the world's most dangerous terrorists, Al Qaeda leader Osama bin Laden, who had been living in the Sudan. Under pressure from the U.S. and the Saudis, the Sudanese government asked bin Laden to leave. In doing so, the Sudan, which wanted to resume normal relations with the U.S., was essentially inviting the Clinton administration to take bin Laden into custody.

As was his customary practice on matters both foreign and domestic—particularly where matters of law were concerned—Bill consulted his wife. Hillary agreed with the President's advisers that, since bin Laden had not yet committed a crime against America, they had no legal grounds for detaining the leader of Al Qaeda. "I said don't bring him here," Clinton admitted years later, "because we had no basis on which to hold him, though we knew he *wanted* to commit crimes against America.

"So I pleaded with the Saudis to take him," Clinton went on, "because they could have. But they thought it was a hot potato and they didn't and that's how he wound up in Afghanistan."

That same month, Hillary traveled to South Africa for Nelson Mandela's inauguration. A few weeks later, she accompanied Bill to England for ceremonies commemorating the fiftieth anniversary of D-day.

Hillary launched an invasion of her own in July, barnstorming the country from coast to coast aboard a bus christened the "Health Security Express." The tour was aimed at whipping up enough grassroots support to convince Congress to reconsider her health care package. "When these guys see the people out there demanding reform," she said, "then they'll get off their asses and do something about it."

Unfortunately, Hillary wildly misjudged the mood of the American people—and how they felt about her. Thousands of demonstrators showed up at every stop to scream obscenities at the First Lady. In Seattle, angry protesters swarmed her motorcade, rocking her limousine and pounding their fists on the windows. Fearing for her life, Hillary agreed for the first time to wear a bulletproof vest.

Once back in Washington, Hillary threw up her hands and admitted defeat. Her well-intentioned attempt to provide universal health care coverage had collapsed under the weight of her own overbearing style. In addition to alienating many in the health care and insurance industries, she had run afoul of leaders in both political parties. "I knew," she later conceded, "that I had contributed to our failure."

Yet the price for pushing her ambitious agenda as "co-President" was to be far higher than the collapse of the Clinton health care initiative. In November 1994 the GOP, led by Newt Gingrich, recaptured the House for the first time since the Eisenhower administration. Republicans also took control of the Senate and most of the nation's governorships. Of these, none wounded Hillary more than the decisive defeat of incumbent Texas Governor Ann Richards by George W. Bush. "God," Hillary muttered as she sat at the kitchen table watching Bush's face flash across the television screen. "What a jerk."

The success of Gingrich's "Contract with America" was a stunning rebuke of the Clinton presidency in general and Hillary in particular. Blaming herself for the Democrats' defeat at the polls, Hillary sank into a deep depression. Rather than run top-level policy meetings at the White House, she opted out entirely. She told Dick Morris that she no longer trusted her own judgment and that she felt "lost."

Every week, ten high-powered Democratic women would get together in what came to be known as "Chix meetings" to talk over strategies and policy matters. The Chix, including Maggie Williams, consultant Mandy Grunwald, and Susan Thomases, huddled with Hillary in the White House Map Room—appropriately enough—to map out a strategy for her future.

Hillary's lip trembled as she apologized to the Chix one by one for letting them and the party down. It was time, she suggested, for her to retreat from public life. Not surprisingly, the Chix rallied to their leader's side, insisting that she was a role model for millions of women, and that she owed it to them not to admit defeat. "We all felt," one of the Chix later said of that meeting, "that Hillary was the one who should have been sitting in the Oval Office, and that someday she would be. But at that point she was being assailed from so many angles that she just wanted to fold up her tent. We gave her the pep talk to keep her in the game."

Hillary had a role model of her own, and she consulted her with some regularity. With the encouragement of her longtime friend the flamboyant Jungian psychologist Jean Houston, Hillary often sat in her room and launched into long and rambling—albeit decidedly one-sided—conversations with the ghost of Eleanor Roosevelt. As far as her current crisis of conscience was concerned, Hillary imagined that Eleanor would simply have told her to "buck up and carry on."

Neither the Chix nor Eleanor provided Hillary with all the guidance she needed. In typical Hillary Clinton decision-making fashion, she consulted everyone from New Age gurus Tony Robbins and Marianne Williams to her old Park Ridge, Illinois, youth minister Don Jones before deciding how to go about redefining her role.

In the wake of the Republican midterm election sweep, Hillary decided to return to the issue of children's rights—and use it to take potshots at the GOP. When Newt Gingrich suggested that the children of some welfare mothers would be better off in orphanages, Hillary blasted the idea in a speech before the New York Women's Agenda and then in a lengthy article in *Newsweek*. Turning the tables on Newt, she blasted his defense of orphanages as "big-government interference into the lives of citizens at its worst."

As with everything that appeared in print under her byline, it is highly doubtful that Hillary actually wrote the *Newsweek* piece.

According to former staff members, presidential speechwriters were always called in to craft serviceable articles for Hillary. Still, she would say that, with the publication of the antiorphanage piece, "I had found my voice."

Once again, Hillary borrowed a page from Eleanor Roosevelt and started writing a nationally syndicated weekly column patterned on Eleanor's "My Day." Methodical as ever, she summoned a number of bestselling authors to the White House and picked their brains concerning the best way to go about writing a book—something she had never attempted before.

Shrewdly designed to recast Hillary's image as a caring wife and mother—as opposed to a shrill and humorless policy wonk—*It Takes a Village* was a collection of wryly amusing anecdotes and homespun advice interwoven with the First Lady's thoughts on child welfare. Hillary's utopia, as described in the book, was one in which the state functioned as a third parent for every child, poised to step in at frequent intervals throughout that child's life. *It Takes a Village* not only became a bestseller—the proceeds were donated to charity—but was also a giant step away from the brittle, autocratic co-President of old.

The nation may have started warming up to the First Lady, but there was a decided chill between Hillary and the ghostwriter she hired to actually do the work. Hillary would brag that she had written the 320-page book in longhand on yellow legal pads. While she certainly did some work on the manuscript, it was actually Georgetown University journalism professor Barbara Feinman who worked feverishly to complete the manuscript on time.

Before she could finish up, however, someone reported to Hillary that Feinman had been talking to the press. Enraged, Hillary tried to block the last $30,000 installment owed Feinman, who had been counting on the money to finance the adoption of a Chinese orphan. Eventually, Hillary came through with the final payment—but only after Feinman reportedly threatened to sue.

In the end, neither Feinman nor anyone who helped her in the publication of *It Takes a Village* was mentioned by Hillary in the book's acknowledgments. "All she expected," said Feinman's friend Sally Quinn, "was 'Many thanks to Barbara Feinman, whose tireless efforts were greatly appreciated.' She would have died and gone to heaven."

Hillary bristled when asked if she really wrote the book herself. "All I can say," she answered with a smirk, "is that they didn't pay me $120,000 to spellcheck it."

Basking in the unfamiliar warmth of public affection, Hillary patterned her next literary endeavor after Barbara Bush's hugely successful *Millie's Book*. A collection of letters penned to the Clintons' pets, *Dear Socks, Dear Buddy* also climbed the bestseller lists, and further burnished Hillary's image as a homebody.

In truth, Hillary was finding the traditional role of First Lady more enjoyable than she might have imagined. She threw herself into holiday preparations at the White House, but soon learned that even the mundane-sounding business of decorating the Blue Room Christmas tree—considered the First Lady's tree—was fraught with the potential for scandal.

Hillary had invited art students from around the country to design their own ornaments for the Blue Room tree along a "Twelve Days of Christmas" theme, with rather unexpected results. A number of the ornaments were fashioned from condoms, others from crack pipes. One ornament depicted "twelve lords a-leaping"—all displaying erections.

Perhaps Hillary was too distracted by her pal Webb Hubbell's indictment on tax evasion and mail fraud charges. Within weeks, Hubbell would confess to overbilling Rose Law Firm clients to the tune of nearly $400,000 and be sentenced to twenty-one months in federal prison.

Concerned for Webb's welfare—and perhaps worried that

he might spill some additional information regarding her work for the failed savings and loan Madison Guaranty if he felt abandoned—Hillary breathed a sigh of relief when Clinton insiders told her they were banding together to help Webb out. They arranged for Hubbell to be given work as a consultant that would pay him more than $400,000 in fees—"enough," said a partner in one of the firms, "for him to keep his mouth shut about Hillary."

In prison, Hubbell contemplated countersuing his former employer over the amount they claimed he owed. But when he was told that Hillary would pull the plug on White House support if he went ahead and sued the Rose Law Firm, Hubbell backed down. "So," he told his wife in a phone call from prison, "I need to roll over one more time."

Hillary was having far too good a time in her new incarnation as a somewhat old-fashioned First Lady. To sell herself in the role, however, she relied increasingly on Chelsea. The First Daughter had been shielded from the press in a way that none of her predecessors had. Bill and Hillary would not entertain any inquiries from the press about Chelsea, and reporters knew they would be banished from the White House if they dared to ask Chelsea even one innocuous question. After years in the White House, it seemed nothing short of incredible that Chelsea had never uttered a word for public consumption. The average American had no idea what Bill and Hillary's only child sounded like.

But Chelsea did serve a very important political purpose. Whenever her parents were under the gun, the First Daughter was often carted out to be photographed laughing with her mom or strolling arm in arm with Dad across the White House lawn.

Through it all, Chelsea remained remarkably unspoiled, exhibiting qualities of poise and self-assurance that impressed visiting heads of state and Arkansas good ol' boys alike. In March of 1995, she accompanied her mother on a twelve-day visit to five countries in South Asia. They were photographed laughing as they rode an elephant in Nepal, holding hands at the Taj Mahal, and touring Mother Teresa's orphanage in Calcutta.

Back in Washington, Hillary was scoring a public relations triumph in absentia. Before leaving on her trip, she videotaped a five-minute parody of the film *Forrest Gump* to be shown at the annual Gridiron Dinner. Never one to leave anything to chance, Hillary tapped several of her friends in the entertainment industry to make the short video. *Saturday Night Live* alumnus Al Franken was asked to direct, while Jay Leno enlisted the help of his writers to provide some of the more memorable lines.

Journalists and politicians howled at the video showing Hillary, on a park bench in front of the White House, spoofing some of *Forrest Gump*'s most famous lines. "My mama always told me the White House is like a box of chocolates," she mugged. "It's pretty on the outside, but inside there's lots of nuts." Whenever the camera came back to Hillary sitting on the bench, she was wearing yet another hairstyle—something Americans had become accustomed to as the First Lady tried over the years to find the one most suitable for her.

Hillary returned from her South Asian tour a few days later, blissfully unaware that while she was away her husband had begun his fatal flirtation with a twenty-two-year-old White House intern named Monica Lewinsky.

Buoyed by the success of her first official overseas trip without Bill, Hillary journeyed to Beijing in September of 1995—this time to attend the United Nations Fourth World Conference on Women. It was a trip the President's advisers did not want Hillary to make. They were concerned that it might appear as if the First Lady were once again trying to act as a surrogate for her husband—this time in the foreign policy area—and that Hillary, whose off-the-cuff remarks had caused so much trouble for the President in the past, might slip once again. Most important, however, was the undeniable fact that the Chinese government would exploit Hillary's presence as a tacit endorsement of its human rights policies.

To further complicate matters, the Chinese government had arrested Chinese American activist Harry Wu and charged him with espionage. Human rights groups pressured Hillary to boycott the conference in protest, but she had no intention of missing this golden opportunity to make her debut on the world stage. A week before Hillary was to arrive in China, Wu was tried, sentenced to fifteen years in prison, and deported to the U.S. In return, it was presumed that Hillary might tone down her criticism of the Chinese government's human rights record or forgo mentioning it altogether.

Her hosts, as it turned out, were in for a rude shock. In her speech to the conference, Hillary delivered what amounted to a stinging indictment of Beijing's human rights violations, which included forced abortions and forced sterilization. Hillary was interrupted several times by standing ovations, and when she was finished, thousands in the audience rushed to the stage to congratulate her. To celebrate Hillary's "Women's Rights Are Human Rights" speech, the First Lady joined hands with other members of the U.S. delegation to form a circle, then led them in a rendition of "Kumbaya."

Even before she left for China, the First Lady had begun planning the fund-raising effort that would bankroll her husband's reelection campaign. While Bill's advisers insisted that it was important that she return to keeping a relatively low profile, they did not dispute the fact that no one was more adept at finding ways to raise cash than Mrs. Clinton.

Commerce Secretary Ron Brown was one of those who did not approve of the Clintons' practice of literally selling seats on the Commerce Department's foreign trade missions for $50,000 apiece. Before his death in a plane crash while on a trade mission to Croatia, Brown complained bitterly about the First Lady's involvement in trying to squeeze dollars from any available source—

and her insistence that he accommodate fat-cat contributors. He told his business partner and mistress, Nolanda Hill, that he was tired of being "a motherfucking tour guide for Hillary."

Hillary's greatest fund-raising tool was the Executive Mansion, and the trappings of the presidency itself. On a scale that was heretofore unimaginable, she masterminded the scheme to make everything that the White House had to offer available—for a price. Contributors who had never actually met the Clintons could stay in the Lincoln Bedroom, attend a state dinner, sit in the Oval Office while the President read his weekly radio address, screen a movie in the White House theater, or sit in the presidential suite at the Kennedy Center.

For $10,000, a campaign contributor could join a small group for coffee with the President, or a larger group for dinner. Those who forked over a six-figure contribution were invited to sit at the President's table. At Hillary's urging, the President also gave personal tours to some donors, and went golfing or running with others. Even the compulsively gregarious Clinton occasionally grew tired of the endless "meet-and-greet." Unimpressed, an exasperated Hillary lit into her husband while staffers looked on nervously. "You're getting your ass out there," Hillary said, "and you're doing what has to be done. *We need the money.*"

One of the more distressing examples of Hillary's blatant influence peddling involved the Riadys, Indonesian billionaires who owned the LippoBank of Los Angeles. For a time in the 1980s, James Riady was president of Arkansas's Worthen Bank, and it was in that capacity that he befriended Hillary and Bill. James Riady and his wife gave $465,000 to Bill's 1992 presidential campaign, making them the Clintons' largest single contributor at the time.

Lippo executive John Huang, who escorted Hillary and Bill on a tour of Hong Kong in 1985, would also give generously to the Clintons. Once their friend was elected, Huang and Riady each gave $100,000 to the Clinton inaugural.

This was just the tip of the fund-raising iceberg. Additional money poured into the Democratic coffers from the Riadys' friends, relatives, and business associates. As a result of their largesse, the Riadys and Huang, now their top-ranking U.S. executive, were allowed to roam the halls of the White House almost at will.

In June, Riady and Huang were among guests invited to the President's Saturday radio broadcast, and after it was over they remained behind closed doors with the President. Two days later, a Riady-owned Chinese company issued a $100,000 check to Hillary's embattled friend Webb Hubbell. The same day, Huang was appointed to a sensitive position at Commerce. In that capacity, Huang was given CIA briefings and ready access to top secret intelligence documents—several, a Senate investigation would reveal, would have cost the lives of CIA operatives in China had they been leaked to the wrong people.

Trouble was, the Riadys and Huang had close ties to Beijing and especially to the Chinese intelligence community. By the 1990s, the Riadys had invested more than $8 billion in various Chinese projects, from banking and electronics to real estate and tourism. To protect their interests, they relied heavily on their relationships with several high-level Communist Party officials—many of whom were escorted by Hillary's friend John Huang to the White House for coffee with the President. All in all, Huang would visit the White House more than seventy times during the Clintons' first term—always with Hillary's blessing, if not the Secret Service's.

Huang was not alone. Hillary was also very fond of Southern California businessman Johnny Chung. Often described as a "Hillary groupie," Chung sent a letter of condolence to the First Lady following her father's death, and thereafter was basically given carte blanche to hang around Hillaryland. Chung racked up more than fifty visits to the Executive Mansion, bringing his well-connected Chinese friends to various functions, including a couple of White House Christmas parties.

When Chung asked Hillaryland staffer Evan Ryan if he could bring some friends—all top officials of the People's Republic of China—to meet the First Lady and dine in the White House mess, Chung claimed he was told Hillary needed to raise $80,000 to pay off a debt to the Democratic National Committee.

Chung eagerly volunteered to help, and returned with a check for $50,000. Within hours of receiving the check, a beaming Hillary met with Chung ("Welcome, my good friend") and his Chinese associates.

A few days later, Chung and company showed up to watch the President deliver his Saturday radio address. For his part, Bill was concerned about the obvious presence of Chinese Communist Party officials in the Oval Office—not because they posed a security risk, or even because their presence implied tacit U.S. approval of the ruthless regime in Beijing. The President told Hillary he was worried about "how it's going to make me look" when photos of Chung and his friends posing with the President in the Oval Office were released.

Unlike Al Gore, who got into trouble for raising $166,750—$55,000 of that laundered through monks and nuns—during a John Huang–organized visit to the Hsi Lai Buddhist Temple in California, Hillary kept a relatively low profile. Yet no one doubted that she was the mastermind behind soliciting donations from special interests outside the U.S. "That's just right-wing bullshit," she said when questions of national security were raised. Regardless of where it came from, she reminded the Clintons' inner circle, "money is money."

That January, Hillary once again found herself confronting the ghost of Whitewater when her long-lost Rose Law Firm billing records inexplicably materialized on a table in the White House residence. The records showed that, contrary to Hillary's insistence that she had not been involved with negotiations between the McDougals' Madison Guaranty and state regulators, she had in fact done sixty hours of work on the matter. As for Castle Grande (a.k.a. Whitewater), which she also claimed to know nothing about, the records showed she had racked up thirty billable hours of legal work.

The revelations sent Hillary's point man Harold Ickes, the dyspeptic son and namesake of FDR's interior secretary, over the edge. Ickes was most concerned that Hillary's office and the Arkansas securities commission get their stories straight. "If we fuck this up," he told them, "we're done."

As inconceivable as it seemed, the First Lady of the United States was now fingerprinted so that investigators could determine which documents she had handled and which she hadn't. "That struck me," said Jane Sherburne, one of the attorneys brought in to handle the case, "as another indignity she had to endure, but part of the process."

With the possibility of an indictment still hanging over her head, in January 1996 Hillary was called to testify before the grand jury investigating Whitewater. "Cheerio!" she said to her team of lawyers as she left the White House to testify. "Off to the firing squad!" After answering questions for more than four hours, Hillary stepped outside the courthouse and fielded a few more from the press. The First Lady remained calm throughout the ordeal, but she confided to friends that she found the experience both "demeaning" and "scary."

Hillary resolved to maintain a low profile during her husband's reelection campaign against Kansas Senator Bob Dole. Yet behind the scenes, she pulled the strings of the Clinton reelection effort with Oz-like dexterity. Helping her in this effort was WhoDB, the computer database locked away in the old Executive Office Building that contained detailed information on more than 350,000 people—and illegally obtained secret FBI files on nearly one thousand officials from the Reagan and Bush administrations.

"Big Brother," as the database was known to the few White

House staffers clued in to its existence, was originally conceived by Hillary as the cyber equivalent of the card file the Clintons had always kept on each contact made during their climb to power. Hillary quickly discovered, however, that the same technology could also be tapped to provide the Clintons with a database that would make Nixon's infamous "enemies list" seem laughable.

In the past, the FBI had shared such sensitive and highly personal information—including criminal and medical records, financial information, and reports on sexual activity and preferences—only for security clearance purposes and with the strict understanding that it would remain highly confidential. Toward that end, those White House officials put in charge of handling these personnel files were usually longtime government employees whose ethics were above reproach.

Hillary had something different in mind. Though she would later testify that she did not even know the man, the First Lady insisted that well-known campaign dirty trickster Craig Livingstone be appointed director of the White House Office of Personnel Security. A former bouncer best known for dressing up in a chicken suit and trailing President George H. W. Bush around during the 1992 campaign, Livingstone had complete access to Big Brother and all the potentially damaging information it contained. Hillary, who met with him several times in the family residence, also assigned Livingstone to perform several sensitive tasks—most notably, identifying Vince Foster's body and helping to tidy up Foster's office.

Incredibly, Hillary would brush off Big Brother's existence with two sentences in her memoirs. A "midlevel" staffer had "blundered" by referring to "an outdated list to order FBI file summaries for current staff, and had inadvertently been sent files on some security pass holders from the Reagan and first Bush administrations. But it was neither a conspiracy nor a crime." Filegate, she boasted, "was a dry hole."

Largely unaware of the key role Hillary played behind the scenes as a cunning strategist, the public was still warming to her new, softer, less officious persona. When Bill turned fifty, she showed up at New York's Radio City Music Hall to praise him as the best man she had ever known—and to gloat over the $10 million that one event raked in for the Democrats. After his landslide 1996 election victory, Hillary once again took to the dance floor at each of the fourteen (up from eleven in 1993) inaugural balls. And with each new State of the Union Address, Hillary was there, applauding her husband from the balcony.

Even as she postponed her own ambitions, The Plan was never far from Hillary's mind. "There was a lot of talk about 'The Plan,'" recalled a junior White House staffer. The Clintons "joked around about it in a cloak-and-dagger way, but you could tell they were serious." After Bill Clinton left office, "their entire focus was going to be on getting Hillary back in."

Indeed, there was also talk of how the Clintons might extend their influence well into the twenty-first century. As a child, Hillary was repeatedly told by her mother that she would someday sit on the United States Supreme Court. Assuming she could win back the White House in 2008 or even 2012, Hillary might well be in the position to appoint a chief justice—assuming that Rehnquist, who would turn eighty-four in 2008, stepped down or died during her term in office.

However, having been denied an official position in her husband's administration because of antinepotism laws, Hillary wondered if she would be legally permitted to appoint her husband to the bench. A test case presented itself in late 1995, when Bill appointed William A. Fletcher to be a judge on the U.S. Court of Appeals for the Ninth Circuit, the same court his mother, Judge Betty Fletcher, had served on since 1979.

In an opinion requested by Assistant Attorney General for Policy Development Eleanor Acheson, Hillary's old Wellesley pal, the counsel to the President concluded that the prohibition "does not apply to presidential appointments of judges to the federal judiciary." Without arousing public suspicion, the First Lady had her answer: Hillary could, were she to occupy the White House in the future, appoint Bill to the Supreme Court—ideally to replace Rehnquist, who had already served on the court for thirty years. (Under those circumstances, Bill would not be the first person to have headed two of the three branches of government. After one term as President, William Howard Taft served as chief justice of the Supreme Court for nine years.) Of course, there was no way of knowing at this point—when he seemed to be enjoying unprecedented popularity with the electorate—that the President's future actions would make any such scenario impossible.

For the time being, it was Hillary, not Bill, whose political future hung in the balance. There was a growing fear inside the White House that the First Lady might be indicted as a result of either the Whitewater or Travelgate investigations. Not long after the McDougals were convicted of bank fraud, Bill asked Dick Morris what he thought of offering Hillary a blanket pardon. Having just been handed a mandate by the American people, the President wondered if now wasn't the time to take action to spare Hillary.

Morris replied that such an act would almost certainly be seen as arrogant. He also told Clinton that "if he tried it," he would go down in history alongside Gerald Ford, whose pardon of Richard Nixon cost him reelection.

When Morris phoned the First Lady and asked what she thought of Bill offering her a preemptive pardon, Hillary "flew into a rage," Morris said. If Whitewater Special Prosecutor Kenneth Starr decided to "play that way," she went on, "I will fight it with all that I've got! I don't want any pardon. I won't take any pardon!"

Clearly, Hillary was feeling the pressure. It didn't help that she was also facing up to the realization that she and Bill would soon be losing the single most important person in their lives—the one who defined them as a family: Chelsea. It was doubly hurtful to Hillary that Chelsea chose to put a continent between her and her parents by enrolling in Stanford University.

As crestfallen as she may have been over Chelsea's departure, Hillary scarcely showed it as she prepared to celebrate her fiftieth birthday on October 26, 1997. Widely acclaimed as the hero of middle-aged women everywhere, she appeared on the cover of *U.S. News & World Report* and *Time,* and was the subject of several television specials. Remembering the $10 million take at one of her husband's fiftieth-birthday parties, Hillary made sure she cashed in on several of her own. There was a gala at Washington's Ritz-Carlton Hotel, another at the White House, and yet another in Chicago, where Oprah Winfrey told Hillary—and a nationwide audience of millions—that she had never looked better.

Hillary was indeed radiant—in part because her 60 percent approval rating in the polls was the highest she'd had since 1993. But she also relied increasingly on the experts—most notably her hairstylist Christophe, designer Oscar de la Renta, and *Vogue* editor Anna Wintour—in completing her transformation from fashion frump to sleek urban sophisticate. George Stephanopoulos echoed the sentiments of all who knew her when he observed that never before had Hillary looked so . . . happy.

At the annual White House Christmas party that year, Hillary stood next to her husband in the reception line, smiling and shaking hands with party functionaries, contributors, and the occasional old pal from Arkansas. One of the guests, New York Democratic Party Chairwoman Judith Hope, lingered for a moment to chat with the First Lady. Hope did not think incumbent Democratic Senator Patrick Moynihan was going to run for a fifth term, and she wanted the First Lady to consider replacing him. "A lot of people," she told Hillary, "think when you leave the White House, you ought to run for U.S. senator from New York."

Hillary laughed off the suggestion at the time. But once she returned to New York, Hope quietly championed the idea among party leaders. Not so fast, said incumbent Democratic Senator Patrick Moynihan. Talk of floating the names of possible replacements was premature, the senator said; Moynihan had every intention of serving out the remaining three years of his term before retiring.

The White House pressured Hope not to continue, and Hillary was personally so distraught at the prospect of being at the center of another controversy that she personally asked the New York State Democratic chair not to mention the idea again. Although she later claimed that at this point she thought the idea was "far-fetched" and even "absurd," Hillary was already on the case. She was, however, unhappy that Hope was tipping her hand too early in the game.

There was another catch: John F. Kennedy Jr. had approached Hope earlier in the year and told her *he* was interested in running for Moynihan's seat in 2000. John hesitated, concerned that his hypersensitive wife, Carolyn Bessette Kennedy, might not be able to hold up under the strain of a political campaign. But for a time it would remain uncertain whether he would throw his hat in the ring.

Hillary was fond of John Kennedy, and like many women somewhat in awe of his devastating charm and matinee-idol looks. She also appreciated the hard political reality that Kennedy, should he choose to seek the Democratic nomination, would be difficult to beat. Not only was he both heir to the Kennedy magic and *People* magazine's "Sexiest Man Alive," but John was the consummate

New Yorker, a resident of the city since the age of three. If she did decide to run, Hillary, who had never spent more than a few days at a time in Manhattan, would be branded a carpetbagger. While he had yet to run for office, it was clear to anyone who had heard JFK Jr. give a speech or field reporters' questions with grace and humor that he was a natural politician.

For Hillary, there were other, more personal reasons not to go up against JFK Jr. for the New York Senate nomination. Beginning with Bill's prescient handshake with JFK in the Rose Garden and running through their friendship with Jackie, Hillary felt a special attachment to the Kennedys—especially to John and to Caroline, who had become a confidante of Chelsea's.

"She was really torn," said one of the few friends Hillary had from New York. "She liked John and hated the idea of running against him, and she also felt he would be impossible to beat on his home turf. It was Bobby Kennedy's seat, and she felt if he wanted it, John should have it." If John Kennedy decided to run, Hillary was now saying, then she would not.

John Kennedy's possible entry into the Senate race hung over Hillary's head for months—ending only with his untimely death in July 1999. Until that point, said a friend, "she was always looking over her shoulder, a little worried that he might change his mind."

There would be plenty of other distractions in the meantime. Just a few weeks after Judith Hope uttered the first quasi-public words regarding a Hillary Clinton for Senate campaign, Bill Clinton gave a six-hour deposition in the Paula Jones case. For the first time, he was asked under oath if he had ever had sexual relations with an intern named Monica Lewinsky—an accusation he flatly denied. Then he rushed home to tell Hillary his version of events: that his secretary, Betty Currie, had this friend who was going through a rough patch personally, and that he tried to console the young woman on a few occasions. In the course of those chats, he claimed, this mentally unbalanced intern—her friends called her "The Stalker," Bill said—somehow began imagining that they had some sort of relationship. . . .

Just like all the other women, Hillary thought to herself. A tramp. Not just a tramp, but a delusional tramp. This time, Hillary would have to work hard at convincing herself of this. The Clintons canceled their plans to dine out, and spent the weekend, she later said without a trace of irony, "cleaning out closets."

On January 21, Hillary woke to Bill sitting on the edge of her bed. In his hand was a copy of that morning's *Washington Post* with a front-page story about the President's alleged affair with Lewinsky—and charges that he had urged her to lie about the affair to Paula Jones's lawyers. Hillary went ballistic, but quickly pulled herself together. Swinging into action as the President's chief strategist, she told Bill he would have to ignore the screaming headlines and go ahead as planned with his schedule.

Hillary would do the same, boarding a train for Baltimore, where she was scheduled to speak at Goucher College. During the ride, the President would call her on her cell phone three times—and each time she would refuse to take the call. Later, when it was revealed that Bill had given Monica a number of gifts, including a copy of Walt Whitman's *Leaves of Grass,* Hillary became noticeably upset. "He gave me the same book," she told an aide, "on our second date."

Still, Hillary stood beside him and nodded enthusiastically as her husband delivered his famous, finger-wagging "I did not have sex with *that woman*" denial. The following morning, Hillary was equally forceful as she backed up her husband on the *Today* show. As she had so many times before, Hillary refrained from assigning any blame to Bill. When interviewer Matt Lauer refused to let her get away with characterizing the scandal as the result of "rumor and innuendo," she pointed an accusing finger at "the vast right-wing conspiracy that has been conspiring against my husband since the day he announced for president." Hillary would later tell White House lawyer David Kendall that, while she was fishing for just the right words to say in her husband's defense, two simple words kept running through her mind: "Screw 'em!"

Politics is conflict.
So is marriage.

—Hillary

▪

Never get a divorce.
Endure everything.

—Dorothy Rodham,
Hillary's mother

▪

I have to kick his ass every morning.

—Hillary, on Bill

She could cut your heart out with her tongue.

—Thomas Mars, a colleague at Hillary's Little Rock law firm

father. Now they asked her to fly to Washington for a show of family unity at Camp David. As convincing as Chelsea's performance was, Hillary knew that she was having a hard time coping with the seamy headlines. The First Lady told Chelsea to follow her lead and simply stop reading the newspapers.

Chelsea took her mother's advice, but it would soon become impossible for anyone to avoid the torrent of sordid details. Starting in May, Chelsea was rushed to the Stanford campus hospital no fewer than four times with stress-induced stomach pains.

Hillary tried to keep up her daughter's spirits with daily phone calls—at this point neither woman was talking to Dad—but she was finding it increasingly difficult to keep up her own. There was the occasional bright spot: On April 1, while the Clintons were in Senegal as part of a twelve-day tour of Africa, U.S. District Court Judge Susan Webber Wright dismissed Paula Jones's case against the President on the grounds that Jones could not prove she had suffered any actual harm even if she had been sexually harassed by Clinton.

Hillary cheered the news, and she and her husband celebrated the surprise decision with several staffers by polishing off several bottles of Dom Pérignon in the Clintons' hotel suite. Then Hillary, whose Sunday-morning talk show warriors had done so much to portray Paula Jones as lying "trailer trash," smiled as Bill lit up a cigar and started pounding on a drum that had been given to him by an African chief.

In the end, the Clintons would pay Jones $850,000 to keep her from appealing the decision. No matter. The Lewinsky scandal had long since taken on a life of its own, spewing out one squalid image after another as it spun out of control that spring and summer of 1998. Among them: the semen-stained navy blue dress; oral sex in the Oval Office (once while Clinton talked on the phone with a congressman about deploying U.S. troops in Bosnia); phone sex; thong underwear; a cigar used by Bill to violate Monica. Then

5

Even after she learned that Bill had been lying to her an
the nation all along, Hillary would never abandon the
tion that nefarious forces were at work to destroy her and
husband—"an interlocking network of groups and individu
she wrote in 2003, "who want to turn the clock back on man
the advances our country has made." These forces, she contin
use "money, power, influence, media, and politics to achieve
ends." These were the true masters, she insisted, of the "politi
personal destruction." Certainly Hillary, who had so skill
masterminded the smear campaigns directed at her husband'
male accusers, knew something about the politics of personal
struction.

If anything, Hillary's appearance on *Today*—and the "ri
wing conspiracy" comment—poured gasoline on the fire. M
than ever, she was concerned about the effect all this was ha
on Chelsea. Hillary called her daughter several times at Stan
warning Chelsea not to pay any attention to the stories abou

there were Lewinsky's nicknames for Bill: "The Creep" and "The Big Creep." Hillary would later learn that Monica had a nickname for her, too: "Baba," an abbreviated form of the Russian *babushka*.

"I've got to believe my husband," Hillary kept telling her friends and her husband's lawyer, David Kendall. "I've got to believe him. . . . He's done lots of lousy things, but," she added with a straight face, "he has never lied to me." Hillary did what she always did in situations like this: she channeled whatever rage she may have felt over this all-too-familiar predicament not at Bill, but at her husband's accusers—and the press.

That would abruptly end on August 13, 1998, when Hillary was again stirred awake by Bill. Rather than sit next to her on the bed, Bill kept a safe distance, pacing the room as he confessed that he had "done some things I shouldn't have" with Lewinsky. Not sex per se, he continued to insist, hewing to his definition that sex meant only intercourse. But "intimacy" that was "inappropriate."

Hillary began hyperventilating, much as she had done that time back in Arkansas—the night when she was rushed to the emergency room with a panic attack. This time, she screamed obscenities at Bill at the top of her lungs. "What do you mean?" she yelled as Bill turned vermilion. "What are you saying? Why did you lie to me? You stupid, stupid, stupid bastard!" Hillary leaped up and slapped Bill hard across the face.

Bill's chronic infidelity—he would confess to Dick Morris that his sexual addiction had led him into relations with hundreds of women—was something Hillary had always been able to "box off." But so long as their political partnership was to work, she counted on her husband to be candid with her when they faced a genuine crisis. Hillary didn't know if their marriage "could—or should—survive such a stinging betrayal." It was, she allowed, "the most devastating, shocking and hurtful experience of my life."

That Sunday, on the eve of Bill's grand jury testimony, Hillary summoned Jesse Jackson to the White House residence to have a heart-to-heart talk with Chelsea. Both women hugged Jackson and listened to him talk about the great biblical figures who had succumbed to temptations of the flesh. Taking Chelsea aside, the Reverend Jackson reassured her that God forgave men "in their weakness," and that the Clinton family needed healing from the Lord.

Conveniently, Reverend Jackson neglected to mention that, at the very time he was giving spiritual guidance to the Clintons, he was having an extramarital affair with a young woman on his staff named Karin Stanford. Months later, Stanford ("From what I understood about Rev. Jackson's marriage is that it was basically a political marriage") would give birth to Jackson's child, touching off a scandal in his ministry and forcing the reverend to make some heartfelt apologies of his own.

Still furious at her husband and—more to the point—unwilling to hear the sordid details, Hillary stayed upstairs in the family residence while Bill was in the Map Room giving his grand jury testimony over closed-circuit television. Although she still wasn't talking to him, when Bill emerged from the Map Room four hours later, she felt a twinge of sympathy. He looked both spent and furious. Realizing that it could not have gone well, Hillary instructed Bill's staff to call Democratic leaders in Congress and reassure them that it had.

Hillary never made a secret of the fact that she, more than anyone, loathed Ken Starr. At a strategy session held in the White House solarium, Paul Begala, James Carville, the Thomasons, and a few others gathered to determine what the President should say that night in his televised apology to the nation. Hillary broke her silence long enough to say that she wanted him to go after Ken Starr. But others—including Give 'em Hell Carville—cautioned him to be contrite and not appear combative. When Bill turned to Hillary to ask her what she thought, she pushed back her chair and got up to leave. "Well, Bill, this is your speech," she said. "You're the one who got yourself into this mess, and only you can decide what to say about it."

In the days before her husband came clean, Hillary had already helped put into motion one series of events conveniently timed to distract the public from the Lewinsky affair and at the same time make her husband look, well, presidential. She had already pointed out that Saddam Hussein's refusal to allow UN weapons inspectors into Iraq might require military action. But then, just two weeks after Al Qaeda's bloody bombing attack on U.S. embassies in Kenya and Tanzania, intelligence reports put Osama bin Laden and his deputies at a training camp in Afghanistan. Hillary concurred with Bill's foreign policy team that now was the time to strike.

The convenient timing of the missile strike seemed truly remarkable, especially since, on several other occasions when U.S. intelligence had pinpointed bin Laden's whereabouts, no action had been taken. "It struck a lot of us as odd," says a retired Pentagon official, "that the President had suddenly awakened to the threat of terrorism and was willing to take bold military action. Word filtered down that Mrs. Clinton kept saying that her husband should do something because 'the President of the United States should not appear weak to the rest of the world.' That wasn't exactly consistent with what she'd been saying for the last thirty years."

Just hours after the President attacked Ken Starr—much to Hillary's delight—in his noticeably less-than-penitent four-minute speech to the nation, missiles were launched on the Al Qaeda training camp where Bin Laden was supposed to be hiding. But by then, bin Laden had gotten word of the assassination attempt and moved on.

In her zeal to launch a counterattack on Starr, Hillary had badly misjudged the mood of the nation. On the Hill, Republicans and Democrats instantly voiced their disappointment over the President's failure to show remorse. Even former members of his own administration—most notably Stephanopoulos (who had left to join ABC News in January of 1997) and ex–Presidential Press Secretary Dee Dee Myers—expressed regret that Clinton had blown "this one chance," as Myers put it, "to make things right."

Before they left the next day for their previously scheduled vacation in Martha's Vineyard, Bill sent a message to Hillaryland asking if the First Lady would issue a statement to the effect that she had forgiven him. Barely able to contain her rage, Hillary made Bill wait until the next morning for her answer. In a statement that she had worked on for hours, Hillary insisted that she was committed to her marriage, and described her love of Bill as "compassionate and steadfast."

At the same time, Hillary made it clear that she, too, had been led astray by Bill—and that, in previously defending him, she had never knowingly lied to the American people. It was a stance that, for the moment at least, she had to take—if for no other reason than to preserve her own credibility. "Bill may not have liked it," said a law school classmate and frequent White House guest, "but if he was going down, she had no intention of going down with him. Hillary wanted to play a role in public life after they left the White House, and I know it was always in the back of both their minds that she would run for President down the line." As Hillary herself put it, "Right now I don't know if he has a future, but I intend to."

It was no small irony that, as a direct result of her husband's atrocious behavior, Hillary's standing in the polls soared. She was the ultimate wronged woman, exhibiting a kind of grace under pressure that surprised even her harshest critics.

Not that she wasn't hurting—and deeply. In an obvious attempt to make small talk at a Vineyard party, CBS newsman Mike Wallace asked Hillary if she had ever had a stress test. "I'm having one now," she answered.

"Her stoic exterior," Dick Morris observed, "masks enormous pain." This latest act of betrayal had left her deeply scarred emotionally. But it was also true, Morris would point out, that Hillary "is never happier than when she can rescue him." The reason: "Because then Bill invests her with even more authority, and the balance of power shifts in her favor."

If, as she claims, Hillary did a great deal of soul-searching before deciding to forgive her husband, it was certainly accomplished with lightning speed. Two days day after he confessed that he had been lying to her and the nation all along about his relationship with Lewinsky, Hillary had set in motion a plan for Bill to seek divine guidance in his quest for redemption—and in the process send a message to the electorate that he had seen the error of his ways. Hillary's three-man "God Squad" (which included Gordon MacDonald, who had been tossed out of his ministry after having an affair with a member of the congregation) arrived at the White House each week to make a show of praying and reading Scripture with the President.

As calls for the President's impeachment grew louder, Hillary found the rationale that would allow her to reenter the fight. "As his wife, I wanted to wring Bill's neck," Hillary later recalled. "But he was not only my husband, he was also my President. . . ."

Just as she had buried her head in the sand by not reading published accounts of her husband's infidelities, Hillary also refused to read the Starr Report when it was released in September 1998. Still, she labeled the 110,000-word report—as opposed to the shocking behavior it described in cringe-making detail—a "low point in American history."

Despite Hillary's professed "compassionate and steadfast" love for Bill, the Clintons kept their distance from each other on their long-planned trips to Russia and Ireland that September. "She couldn't even look at him, much less hold his hand or whisper in his ear," said a reporter who covered both trips. "The tension between them was palpable."

There was also considerable speculation that Hillary was intent on protecting her own political future by distancing herself from Bill—a process that would, one assumed, culminate in divorce. But Hillary soon realized that she was getting more mileage out of being the slighted partner than she ever got out of being a full one.

She may not have read the Starr Report, but Hillary paid close attention to what they were saying about *her* in the press. And she liked what she read. The outpouring of public sympathy for her predicament came not only from Democrats but from conservative Republicans, who now viewed her as the hapless victim of an unscrupulous cad.

Sympathy was one thing. Pity was quite another. Determined not to appear defeated, Hillary jumped at the chance to pose for the cover of *Vogue* in an elegant burgundy Oscar de la Renta gown. At about the same time, she was working out a way to capitalize on her newfound popularity.

With the threat of impeachment hanging over their heads like a sword of Damocles, Hillary realized that the 1998 congressional elections would be viewed as a referendum on the Clinton presidency. She also understood that, if Hillary Rodham Clinton was to have a viable political future, she would have to accumulate some IOUs from Democratic candidates across the country. While Bill agreed to stay put in Washington, Hillary hit the road for the Democrats.

For the next several weeks, Hillary campaigned exhaustively for Democratic candidates across the country while, for the most part, the President holed up in the White House. Hillary's daring election strategy worked. Democrats held their ground in the Senate, and in the House, where they were expected to lose thirty seats, they actually picked up an additional five.

Just a few days later, Patrick Moynihan announced that he would not seek a fifth term in the Senate. Within hours, Harlem Congressman Charlie Rangel was on the phone to Hillary, urging her to run for Moynihan's seat. Both Rangel and Hillary were well aware that any Democrat would likely be facing one of the most formidable candidates imaginable: New York City's phenomenally popular mayor, Rudolph Giuliani.

While she good-naturedly brushed off Rangel's suggestion that she run ("We have a few other outstanding matters to resolve right now"), Hillary had in fact never stopped weighing the pros and cons of a Senate race. "We all knew she wanted it so bad she could taste it," said one state party official. "But she knew it would never happen if President Clinton was run out of office in disgrace." Saving Bill would be an important first step in launching herself as an independent political entity.

The Democrats' Hillary-driven 1998 election victory had yielded an unexpected bonus. The First Lady was overjoyed to learn that, as a direct result of the Republicans' stunning defeat, Newt Gingrich was stepping down as speaker and giving up his seat in Congress. For a moment, she allowed herself to think that Gingrich's departure signaled an end to impeachment efforts.

She was wrong. As Congress continued its march toward impeachment, Hillary once again took it upon herself to play Joan of Arc. Strapping on her armor, she went to Capitol Hill and exhorted the party faithful to hold the line against those dark forces that had been trying "from Day One" to run Bill Clinton out of office. Her stirring performance had the desired effect. As the congressmen filed out of the room, each stopping for a moment to embrace America's embattled First Lady, there was a growing realization that Hillary—far more than her husband—was in control of the nation's political future. "That was the first time," said one of those present, "that a lot of us got our first up-close view of how she does business." It was also the first time congressional leaders "started seeing Hillary as presidential material."

Inside her marriage, things were not quite so rosy. Away from the cameras, Hillary barely spoke to her husband. She still acted, Bernie Nussbaum observed, "like someone had died."

There were also reports from White House employees that, once again, Hillary could be heard shrieking at her husband behind closed doors. On a trip to the Middle East just a few days before the scheduled impeachment vote, the President reached for Hillary's hand only to have her shrink from him in disgust.

By this point, impeachment was a foregone conclusion. In what was yet another remarkable *Wag the Dog* coincidence, intelligence experts were supposedly telling the President that now—just as the impeachment debate was about to begin in the House—was precisely the right time to order air strikes on Iraq. This, despite the fact that Saddam Hussein had been defying the UN on weapons inspections for years.

Bill's advisers knew full well that he would be criticized for putting the lives of American pilots at risk just to divert attention from impeachment. But Hillary, now very much back at the fulcrum of both domestic and foreign policy, insisted that her husband seize any opportunity to appear strong and presidential. One adviser felt Hillary seemed "almost too eager" to see her husband give the order to launch an attack.

Reaction, as predicted, was swift. Joel Hefley echoed the sentiments of his fellow Republicans in the House when he blasted the President's use of the military as a "blatant and disgraceful" attempt to distract the nation from what was about to take place in the halls of Congress.

Hillary had expected a backlash, but none so ferocious as this. The President, on the verge of impeachment, could ill afford a further erosion in his credibility. Over the next six months, there would be no fewer than three opportunities to take out Osama bin Laden—in the Afghan city of Kandahar that December, at a hunting camp in western Afghanistan in February 1999, and again in Kandahar three months later. In each case, officials either doubted the accuracy of the intelligence they were getting, feared that civilians would be killed, or both. "We had a round in our chamber," former Democratic Senator Bob Kerrey would later say, "and we didn't use it."

While Hillary did not sit in on Oval Office meetings with the administration's counterterrorist coordinator Richard Clarke, National Security Adviser Sandy Berger, CIA Director George Tenet, and other high-level officials, her influence was palpable. Bill had never tried to conceal the fact that he valued Hillary's opinion on matters both domestic and foreign, and it was no secret that she did not hesitate to give it to him. Countless Clinton staffers over the years had braced themselves as soon as Bill began a sentence with the familiar words "Hillary says" or "Hillary thinks." It was understood that, regardless of the topic at hand, the First Lady's input was to be taken seriously. After all, as the President was also fond of saying, "Hillary is the smartest woman—the smartest *person*—I have ever known."

It was Hillary who pushed her husband for more involvement in South Asia after visiting India in 1995, and who, two years later, publicly condemned Afghanistan's repressive Taliban regime in a speech before the UN. Teaming up with Secretary of State Madeleine Albright, Hillary also pressured her husband to focus on getting rid of the Taliban.

To some extent Bill Clinton had always pursued a "Hillary Says" foreign policy, just as he had employed a "Hillary Says" approach to politics at home. When advisers split on the question of whether or not to launch air strikes against bin Laden in late 1998 and early 1999, it was Hillary who tipped the scales. She warned her husband that, with his presidency hanging in the balance, the political risk was just too great.

On December 19, 1998, the House voted to impeach the President on charges of perjuring himself before the grand jury and obstruction of justice. Following the vote, Hillary and Bill met a delegation of two hundred congressional Democrats in the Rose Garden. The Clintons smiled and waved along with their supporters, doing their best to look as if they did not have a care in the world. Even Senate Democrats were outraged at what West Virginia Senator Robert Byrd called a "display of egregious arrogance."

Hillary sat in on meetings with her husband and his legal team over the course of the five-week trial in the Senate, but she scrupulously avoided watching it on television. On February 12, 1999, the Senate voted 55–45 to acquit the President of perjury and split evenly on obstruction of justice—falling far short in both cases of the two-thirds majority needed to convict.

Even as the senators decided her husband's fate, Hillary was busy contemplating her own. The same day the Senate verdict was handed down, she held a marathon meeting with her old friend and former Clinton adviser Harold Ickes in the White House family quarters. More than a half century earlier, Ickes's father had urged then–First Lady Eleanor Roosevelt to run for the Senate from New York. Eleanor instantly declined, arguing that she could be a more effective catalyst for social change if she remained a free agent. Her role model notwithstanding, Hillary felt the opposite was true in her case. "I want independence," she said. "I want to be judged on my own merits."

Hillary, who had been playing coy with New York Democratic leaders Judith Hope and Charlie Rangel on the subject of running for the Senate, now wanted an expert opinion. What were her chances, she asked Ickes, of actually winning?

Hillary trusted the abrasive and congenitally crafty Ickes, a battle-scarred veteran of rough-and-tumble New York politics, to be blunt. Ickes unrolled a map of the state and spread it out on a table. Like a general explaining his battle plan, he outlined the difficulties Hillary the candidate would face in taking her case to 19 million New Yorkers spread out over fifty-four thousand square miles. From Syracuse, Buffalo, Albany, and Rochester to the north, to downstate New York City and its sprawling suburbs, Hillary would first have to familiarize herself with local issues and personalities. Even without a rival as formidable as the homegrown Mayor Giuliani, Hillary would face the daunting task of dealing with New York City's powerful political factions and interest groups.

When the President peeked in to say hello, Hillary did not invite him to stay. "The most difficult decisions in my life were to stay married to Bill," Hillary later said, "and to run for the Senate from New York." These were her decisions to make, and while The Plan was still very much in effect, this was a decision she had to make alone.

Bill Clinton didn't need to be asked his opinion about Hillary getting into the race. "The President was right in there, cheering her on—before she even knew she was on the team," Charles Rangel said. "He was the one who asked the most questions about how she could win. You could see the guilt written all over his face. Any man would do anything to get out of the doghouse he was in."

Neither Ickes nor Hillary spent much time debating the obvious: that she was not a New Yorker, and had never spent more than a few days there at a time. "I have been one of the three or four New Yorkers who she has known in her life," Dick Morris said. "Never, never did she comment on the state of the city to me, ask any questions about it, or show any interest in its people, culture, or politics. Not once."

No matter. New York Democrats, desperate to hold on to Moynihan's Senate seat, were scrambling for a candidate strong enough to take on Giuliani. Moreover, New York's residency requirements were virtually nonexistent, and the state had a history of welcoming carpetbaggers—especially famous ones.

Four days after the Senate vote acquitting her husband, Hillary had her office issue a statement acknowledging that she was considering a Senate run. She had, in fact, already launched her campaign effort. Phone in hand, Hillary ran down the list of one hundred well-connected—and mostly wealthy—New Yorkers prepared for her by Ickes. Lining up support from former Democratic mayors Ed Koch and David Dinkins, as well as both Moynihan and the state's other Democratic U.S. senator, Democrat Chuck Schumer, Hillary then summoned Clinton fund-raising impresario Terry McAuliffe to the White House to talk cold, hard cash. It would take $25 million to mount a winning campaign, he told her. "No problem," she replied.

Nor was there likely to be any competition. The only other Democrat who had expressed interest in running, Congresswoman Nita Lowey, would quickly withdraw once Hillary made known her true intentions. And the only other Democrat who could possibly cause her trouble, John Kennedy Jr., was now telling friends that if he ran for office, it would probably be for governor of New York. Still, Hillary actively worried about young Kennedy, seeking assurances from state party officials that he would not be a last-minute entry into the race. "People love John," she conceded, "more than they love me."

Hillary would maintain her high profile as the President's emissary abroad, making a state visit to North Africa in March. Back home, Bill was under fire again—this time over the rape of Juanita Broaddrick twenty-one years earlier. Once again, Serb dictator Slobodan Milošević presented the President with an opportunity to use the American military for purposes of distraction. "I urged him to bomb," Hillary admitted, comparing Milošević's attempts at ethnic cleansing with the Holocaust. "You cannot let this go on. . . ."

Meanwhile, Bill kept nudging his wife in the direction of running. "He wants to run my constituency," Hillary said. "Meet the people and all." But she made it clear that she wanted him to "stay behind the scenes" while she laid the groundwork for a campaign on her own terms. "The tables," she said, "were turned now, as he played for me the role I had always performed for him."

Clinton was also willing to use the presidential pulpit to launch not-so-thinly-veiled attacks against his wife's likely opponent in the Senate race. In early 1999, four undercover New York City police officers gunned down unarmed African immigrant Amadou Diallo when they mistakenly believed he was reaching for a gun instead of his wallet. While Mayor Giuliani urged citizens not to rush to judgment in the case, Hillary asked her husband to devote his weekly radio address to the issue of police brutality. The President told his audience that he was "deeply disturbed" by "recent allegations of serious police misconduct and . . . racial profiling that have shaken some communities' faith in the police." Bill then announced an $87 million federal program aimed at encouraging police recruitment of minorities and putting an end to racial profiling.

To further establish her foreign policy credentials with New York voters and remind them that she had in effect been co-President for eight years, Hillary toured Albanian refugee camps in Macedonia. One of the camps she visited, Brazde, was just ten miles from the Kosovo border—within easy range of Orkan ("Hurricane"), a Serbian weapons system developed with the assistance of Iraq's Saddam Hussein. Orkan was capable of firing hundreds of small rockets with deadly accuracy. While Hillary walked among the refugees who were victims of Slobodan Milošević's "ethnic cleansing," she repeatedly denounced the Yugoslav dictator as "evil."

According to Yugoslav Army Chief of Staff Nebojsa Pavkovic, Hillary narrowly escaped assassination during the visit. Orkan was aimed at the camp, and field commanders awaited Milošević's orders to use it. They never came.

However, just weeks before, Milošević did order Pavkovic to assassinate Tony Blair by blowing up the British prime minister's Puma helicopter at Petrovac Airport in the Macedonian capital of Skopje. Fearing massive retaliation, Pavkovic did not carry out Milošević's orders.

Two months later and light-years away from war-ravaged Kosovo, Hillary chose Patrick Moynihan's bucolic nine-hundred-acre farm in Pindars Corners to announce that she was forming an official campaign committee. This was tantamount to a formal declaration of her candidacy. Senator Moynihan was there, along with his wife and more than two hundred reporters from around the world. Bill was conspicuously absent. He had wanted to come, but Hillary said no.

That same day Hillary, who had alienated many in New York's large Jewish community by calling for an independent Palestinian state, wrote a letter to Orthodox Jewish leader Mandell Ganchow declaring her belief that Jerusalem was "the eternal and indivisible capital of Israel." She also stated her support for relocating the U.S. embassy from Tel Aviv to the Holy City.

Scarcely a week after announcing her candidacy, Hillary was with Bill at Camp David when aides informed them that Hillary's only potential rival for the Senate nomination had vanished over the Atlantic. Days later, John Kennedy's Piper Saratoga was lifted from the bottom of the sea, along with his body and those of his wife, Carolyn Bessette, and Bessette's sister Lauren. Hillary, who had been among the mourners at Jacqueline Onassis's funeral, now attended the memorial service for Jackie's only son at New York's St. Thomas More Church—accompanied this time by the President and Chelsea.

Now the Democrats' only viable candidate for Moynihan's Senate seat, Hillary was soon off on another of her "listening tours" to drum up support. Before embarking, the First Lady was spotted ducking into the Park Avenue offices of celebrity plastic surgeon Cap Lesesne, sparking rumors that she was nipped and tucked in preparation for her Senate run. Lesesne would neither confirm nor deny the reports, saying only that he counted a number of noted political figures among his patients.

Hillary, whose office denied the plastic surgery rumors, crisscrossed the state in a Ford conversion van promptly dubbed the "HRC Speedwagon." She was photographed chugalugging coffee with the morning regulars at diners from Niagara Falls to the Finger Lakes, marching down Fifth Avenue in the Puerto Rican Day Parade, visiting a pediatric ward in Rochester, and wolfing down kielbasa at a deli on Manhattan's Lower East Side.

"I knew Hillary would do well on the campaign trail," state Democratic Party chief Judith Hope later said. "But I was really kind of taken aback by just how good she was."

Hillary's approach to connecting with voters was, like everything else she did, methodical. Before meeting a group of contributors, she would ask for thumbnail sketches of each person she was going to be introduced to—including names of family members, where their children were going to school, vocations, interests, and accomplishments. Recalling a get-together of twenty-five contributors who gave a total of $100,000 to Hillary's campaign, Long Island fund-raiser Marsha Laufer said that, having committed her thumbnail sketches to memory, Hillary "put everyone at ease immediately. With each person, she'd break the ice by asking a question about the college their child was going to, or where they'd just been on vacation." Walking up to one guest who happened to be a theoretical physicist, Hillary broke the ice by asking about a recent breakthrough in his field. "As you can imagine," Laufer said, "he was just over the moon."

Yet Hillary was shrewd enough to realize that fund-raising dinners and photo ops were not enough. In closed-door meetings at the White House, she and her advisers brainstormed about ways to harness the power of the presidency to win over New York's all-important racial and ethnic voting blocs. Any actions on Hillary's behalf would have the added benefit of aiding Vice President Al Gore in his run for the White House.

Bill was eager to do whatever he could to help. "Our entire lives, she has always been there for me," he said. "I owe everything to her." Now, added one FOB, "it was payback time."

At Hillary's urging, the President stunned Justice Department officials by granting clemency to sixteen Puerto Rican terrorists who had been sentenced to prison following a wave of bombings from 1974 to 1983 that took the lives of six Americans and wounded scores of others. Incredibly, the terrorists had not even asked for clemency.

The bloodiest of these attacks took place at Manhattan's historic Fraunces Tavern, where in 1783 George Washington bid an emotional farewell to his officers. On January 24, 1975, a bomb planted by the militant Puerto Rican nationalist Group FALN, the Fuerzas Armadas de Liberación Nacional (Armed Forces of National Liberation), exploded at the height of lunch hour, hurling body parts into the street and killing four people.

When her husband handed out pardons to the FALN leaders, Hillary kept a low profile, careful not to appear as if she had been involved in the process. But neither she nor Bill had anticipated that the terrorists, unwilling to apologize or express any remorse for what they had done, would initially balk at accepting the clemency offer. They feared that, by accepting a presidential pardon, they would be giving up their cherished status as martyrs to the cause.

Over the next three weeks, Hillary pressed her husband to find some way to convince the terrorists to change their minds. The delay, she complained to a longtime Arkansas confidante, "really looks bad." Once satisfied that they had made their point, the FALN leaders accepted the President's offer and walked out of prison to the cheers of their supporters—all the while insisting that they had no regrets for what they had done.

Former U.S. Attorney Joseph DiGenova was among those who branded the FALN pardons as "despicable. . . . Let me say categorically, the Puerto Rican terrorists were pardoned because they were a political benefit to the president's wife. Make no mistake about it."

No one was more outraged than Joe Connor, whose father was killed in the Fraunces Tavern bombing. "My dad . . . didn't have any qualms with the Puerto Rican people. He was just a working guy. He was eating lunch with friends and his life was valued less than that of the president's wife and Al Gore. It's disgusting . . . They believed that by giving clemency this would rally the Puerto Rican vote for Hillary in New York."

Connor, who along with other victims' relatives was not informed of the terrorists' impending release, also made a chillingly prophetic remark at the time—nearly two years before September 11. "The world is a much less safe place," he said, "and this country is a much less safe place, as a result of letting these people out. Certainly, other terrorists might be thinking about attacking us. It will send the wrong message to people who may be planning something."

FBI Director Louis Freeh opposed the pardons, as did the Justice Department, New York Mayor Rudy Giuliani, and even New York's other Democratic senator, Charles Schumer. So did Puerto Rico's popular former governor, Carlos Romero-Barcelo, who pleaded with the President not to release the bombers. "How can we responsibly set them free?" he asked. "What if they kill somebody else? What do we say, 'Too bad'?"

Initially, Hillary claimed she had not been involved in the President's decision and was not familiar enough with the case to comment on it. However, she did express support for her husband's actions. Hillary had understood that the late John Cardinal O'Connor had supported clemency for the Puerto Rican terrorists, and she told her husband's advisers to use that information to defend the pardons. No sooner did the White House float that idea than it was shot down by the Archdiocese of New York, which claimed the cardinal had never supported clemency. In the end, the Senate would vote 95–2 to condemn the FALN pardons.

As public outrage grew, Hillary did an abrupt about-face, claiming that she now opposed the pardons. By having it both ways, Hillary momentarily silenced her critics while at the same time leaving no doubt in the minds of New York's Puerto Rican community that Hillary had done something for them.

Until now, the Clintons had made little use of the President's clemency powers. Over the first six years of their administration, Bill had granted precisely the same number of pardons—74—doled out by his immediate predecessor in a single term. The FALN pardons, however, opened the Clintons' eyes to the ways in which this particular constitutional power could be used to achieve a political—and in Hillary's case, personal—advantage. This President, unlike all who had gone before him, was willing to ignore the Justice Department altogether in the granting of pardons. To the Clintons' unalloyed delight, they could empty the jail cells of their choice and, by merely claiming that it served the interest of justice, suffer no lasting consequences.

In another blatant ploy to curry favor with one of New York's largest voting blocs, Hillary's camp let slip the fact that she had Jewish relatives. Dorothy Rodham's mother, Della, had divorced her husband in 1927 and married Max Rosenberg six years later. Together, they had a daughter—Hillary's aunt Adeline. *The Forward,* a weekly Jewish newspaper, described Hillary's grandmother Della Rosenberg as "the feisty wife of a Yiddish-speaking Jewish immigrant," and suggested that it would enhance her standing with Jewish voters.

Not everyone was convinced, especially since Hillary had never mentioned her Jewish stepgrandfather until this propitious moment. "OY VEY!" proclaimed the *New York Post* on its front page. HILLARY'S *ALMOST* JEWISH. Former New York Mayor Ed Koch, one of Hillary's biggest boosters, said, "I'm a proud member of the

Jewish faith, and it would be wonderful if Hillary were Jewish. But she's not."

The freed FALN terrorists and her newfound Jewish relatives aside, Hillary also had to attend to some rather mundane matters. For purposes of establishing residency in the state, she soon found herself house-hunting. In late August, she and Bill settled on a $1.7 million, 110-year-old, three-story Dutch colonial at 15 Old House Lane in the Westchester County town of Chappaqua, about fifty minutes north of New York City. The five-thousand-square-foot white clapboard house, which boasted a swimming pool and an octagonal screened porch, was situated on a cul-de-sac. While huge Victorian, Federal, pseudo-Tudor, neo-Gothic, and Norman-style mansions loomed on the hill directly behind them, the half-dozen homes on Old House Lane were far from grand. Just a few houses down from the Clintons', there was a bus stop at the intersection of Old House Lane and busy Route 117. Beyond that was Horace Greeley High School, the global headquarters of *Reader's Digest,* and, not far up the road, Mount Kisco Country Club.

After spending eighteen years in government housing—the Governor's Mansion in Little Rock and the White House—Hillary was not quite prepared for the Clintons to take the real estate plunge entirely on their own. They turned to Terry McAuliffe, who was rewarded for his fund-raising wizardry by being named Democratic National Committee chairman, to guarantee their $1.35 million mortgage. McAuliffe could well afford it; he had just reaped an $18 million windfall from a $100,000 investment in the fiber-optics network Global Crossing. After this cozy real estate deal was widely condemned in the press as well as by the Office of Government Ethics, the Clintons got another mortgage—this time sans sugar daddy McAuliffe—with the PNC Bank Corporation.

Her listening tour and the Chappaqua house notwithstanding, many New Yorkers still viewed the First Lady's motives with suspicion. It didn't help matters when well-known Chicago Cubs fan Hillary Clinton posed wearing a Yankees cap right after the team's 1999 World Series victory. "When I saw that," said L. D. Brown, "I was totally disgusted. Hillary was totally, completely devoted to Chicago—the Cubs and the White Sox. She never once mentioned New York fondly at all during the years I knew her in Arkansas, but Hillary *always* talked about how much she loved Chicago, how she wanted to live there someday, the whole bit."

Around the time she proclaimed herself a Yankees fan, Hillary made what could have been the biggest gaff of her campaign during an official visit to the Middle East as First Lady. At one point, she nodded in approval as Suha Arafat, wife of PLO leader Yasser Arafat, accused Israel of, among other things, using poison gas on Palestinians. Then Hillary marched to the podium, embraced Mrs. Arafat warmly—and kissed her.

Hillary would later try to explain that she had not actually understood the translation of Mrs. Arafat's poison-gas remarks, but backpedaled when the translation was released. Clear and precise, it left no room for doubt as to what Mrs. Arafat was saying. Hillary, suggested an Israeli official who was present, "looked as if she was caught up in the moment. It was definitely a very warm exchange between Hillary and Suha Arafat. A lot of us were shocked."

It should have come as a shock to no one, given Hillary's longstanding support of the Palestinian cause in general and the PLO in particular. During the Senate campaign, pains would be taken to downplay the fact that Yasser Arafat had been invited to the White House more times than any other world leader. Hillary also prayed that no one would notice that Arafat was so pleased with the treatment he had received from the Clintons that he had given them $12,000 worth of gold and diamond jewelry—rings, bracelets, and necklaces that Hillary quietly consigned to the National Archives with other offerings from world leaders.

Hillary's overt display of affection toward Arafat's wife was only one of several incidents that aroused suspicion in New York's Jewish community. Later in the campaign Paul Fray, who had managed Bill's first, unsuccessful bid for Congress back in 1974, would confirm stories that Hillary once called him a "fucking Jew bastard." State Trooper Larry Patterson would add that this was nothing new: he often heard Hillary and Bill call each other "Jew bastard" and "Jew motherfucker." In Hillary's case, Patterson said he had heard her make anti-Semitic slurs "at least twenty times" in the heat of anger.

"It did not happen," Hillary responded to Paul Fray's allegation. "I have never said anything like that. Ever. Ever."

In fact, Paul Fray conceded that Hillary's alleged remark "had a lot less to do with religion and a lot more to do with how much Hillary hated me." Five years later, however, Hillary's capacity for making ethnically insensitive remarks surfaced again—at yet another fund-raiser, this time in St. Louis. Managing to both insult one of history's towering figures and reinforce a painful stereotype for many Indian immigrants, Hillary cracked, "You know Mahatma Gandhi. He ran a gas station down in St. Louis." Following the inevitable uproar over the comment, Hillary apologized for what she called a "lame attempt at humor." For those who actually heard her deliver the line, however, it was not just the comment itself. "It was the *way* she said it," commented an audience member. "It was nasty, mean-spirited. And it was done quite naturally, as if she didn't have a clue why anybody would take offense. I was shocked."

Back in 1999, when Paul Fray accused her of calling him a "Jew bastard," Hillary was not about to rely solely on her denials to defuse the scandal. In keeping with her tried-and-true practice of discrediting her accusers, Hillary instructed an aide to send a memo to Clinton's "Jewish Advisory Committee" outlining ways in which to cast doubt on the credibility of the three people who allegedly went on record as hearing Hillary make the anti-Semitic slur: Fray, his wife Mary, and 1974 Clinton for Congress campaign worker Neil McDonald. The same memo urged Hillary's Jewish backers to come forward and claim they had never heard her make an anti-Semitic remark, but not to mention that they had been asked to defend her. Instead, Hillary's memo instructed them to say they were stepping forward on their own because they were "outraged with what was said about her." New York's other Democratic senator, Chuck Schumer, did just that, along with seven other Jewish members of Congress, who within hours of receiving their marching orders from Hillary held a conference at the Democratic National Committee offices to defend her.

"We're talking about something that occurred 26 years ago, but here again," Paul Fray said at the time, "it's indicative of her style of approach that she's going to deny it until it's proven otherwise." Hillary's denials were, in fact, enough to convince voters that she could never have intentionally uttered an ethnic slur or an anti-Semitic remark.

These were the kinds of mistakes lifelong New Yorker Rudy Giuliani was not likely to make. Indeed, Hillary complained bitterly that the New York press was clearly playing favorites, lying in wait for the carpetbagging First Lady to make her next gaff while giving the hometown boy a free pass. Hillary, who in the early months of her campaign restricted press access largely to staged events and carefully circumscribed interviews, also suffered in comparison to the media-savvy Giuliani.

For months CBS talk-show host David Letterman had bombarded Hillary's communications director, Howard Wolfson, with requests for an interview—a campaign that, as time went on, became a running gag on the show. Giuliani had charmed Letterman's audience on several occasions, and the longer Hillary stalled, the more awkward the talk-show host's on-air importuning became. Concerned that she would be "skewered" by Letterman,

Hillary agreed to do the interview only if she knew the questions in advance. To make things even easier, comedy writers provided Hillary with a couple of snappy one-liners. Asking about her new house in Chappaqua, Letterman cracked that "every idiot in the area is going to drive by honking now."

"Oh, was that you?" Hillary shot back. The scripted line left the audience howling. She then went on to astound viewers with her in-depth knowledge of New York State as Letterman quizzed her on the state flower, the state bird, the state tree, the state motto, and more. What viewers did not know, of course, was that Hillary had been given the questions—and the answers—well ahead of time. (In her 2003 memoirs, Hillary would take full credit for the witty retort, and not mention the quiz—or the prepping for it—at all.)

On January 19, Martin Luther King Day, Hillary went to Harlem to speak at the headquarters of the Reverend Al Sharpton's National Action Network. The First Lady–turned–Senate candidate could always count on a warm welcome in New York's black neighborhoods, where her husband was still widely regarded as a great man. Referring to "the tragic murder of Amadou Diallo," Hillary affirmed that "what every community, but particularly the African American community, wants is to be respected and protected. The process has already started to bring this matter to the attention of the federal authorities," she continued, being none too subtle about her willingness to have her husband pressure the Justice Department to intervene. "And if there is a role for the federal government, I will certainly encourage and urge that that occur."

One month later, the four officers—who were never tried for murder—were acquitted on manslaughter charges by a jury that included blacks, whites, and Hispanics. It was only then that Hillary, now coming under heavy fire for unfairly criticizing the police and facing the possibility of being sued by the officers for slander, apologized for blithely calling the four innocent men murderers.

Yet when police killed another unarmed black man in March 2000, Hillary once again journeyed to Harlem to condemn New York City police officers and the mayor who stood behind them. When undercover cops approached Patrick Dorismond outside a bar and asked if he had drugs to sell, Dorismond became angry and a fight ensued. In the struggle, one of the officers shot and killed Dorismond. In a move that could only be described as indefensible, Giuliani, again under siege, released Dorismond's juvenile arrest record. It showed, the mayor said, that the victim was "no altar boy."

Hillary pounced on Giuliani's misguided attempt at defending his beleaguered police department. "I do not believe bad relations with the police are necessary to keep our streets safe," she told 1,100 congregates crammed into Harlem's Bethel A.M.E. Church. "New York has a real problem, and we all know it. All of us, it seems, except the Mayor." Hillary would later savor the moment. "The packed sanctuary," she recalled, "erupted in cheers and hallelujahs." This was, she said, the turning point in her campaign—the moment she gained "traction."

Hillary had wasted no time in playing the race card. While slamming Giuliani for his "divisive rhetoric," she hammered away at what she portrayed as the mayor's callous disregard for the rights and safety of New York's black population. She preferred to ignore the fact that, under Giuliani, New York's crime rate had plummeted. So, too, had police shootings. Under Giuliani's Democratic predecessor, David Dinkins, an African American and a close political ally of Hillary's, there were 212 police shootings in 1993 alone—a number Giuliani had *reduced* by 77 percent. "I didn't see the Justice Department investigating David Dinkins in an election year," Giuliani said, pointing to the fact that as a result of his administration's crackdown on crime, the city's murder rate had dropped from 2,100 a year in the early 1990s to 600 in the year 2000—a decline that was most evident in the city's minority neighborhoods. Instead of investigating New York's Finest, the mayor added, "the Justice Department should be giving them a medal."

A number of Hillary's supporters clearly did not share that sentiment. At the state Democratic Party's convention in Albany that May, they jeered and spat on an Albany police honor guard carrying the American flag. Some of the convention delegates, who would later formally vote to nominate Hillary as their party's candidate for the Senate, pelted the officers with insults. "It's Giuliani's Third Reich!" one protester screamed at the honor guard, which had been invited to participate. Others simply yelled, "Nazis!"

Hillary would later write a letter of apology to Albany Police Chief John Neilsen. She condemned the actions of the few cop-haters in no uncertain terms, but showed no signs of comprehending that her intemperate antipolice rhetoric may have inflamed her supporters.

Over the spring, Hillary crept up in the polls as Giuliani found himself caught up in his own extramarital scandal. While she would later claim to have sympathized with the mayor's plight, at the time Hillary was, said a campaign volunteer, "thrilled about the whole sordid mess. I watched her reading a story about Giuliani's affair in the *Daily News,* and at one point she just *howled* with laughter."

That was nothing compared to the reaction at Clinton headquarters when Giuliani announced on May 19, 2000, that he was withdrawing from the race to battle prostate cancer. Behind the campaign's solemn public face, there was "jubilation," said another Clinton volunteer. "Hillary knew that Giuliani was the only one who stood even the slightest chance of beating her. She didn't want to look as if she was happy he'd gotten cancer, but some of the other people in the campaign made no secret of their hatred for Giuliani."

Hillary said nothing when one of her most trusted advisers reacted to the news with "Cancer? Couldn't have happened to a nicer guy."

Few doubted that Hillary would trounce Giuliani's replacement, a fresh-faced but little-known congressman from Long Island named Rick Lazio. But then Hillary was, in the words of her own mother, "never one to leave things to chance."

Hillary was confident that she enjoyed a significant lead in New York's black and Hispanic communities. But because of tensions between the black and Jewish communities, the kiss she gave Mrs. Arafat, her support for a Palestinian state, and her reputed "Jew bastard" remark, she worried that she might lose New York's significant Jewish vote.

Just to be safe, Hillary made a concerted effort not to be seen in any situation that might make her appear to be sympathetic to the Arab cause. That was going to be tricky, since several individuals and organizations with ties to Arafat, the PLO, and other anti-Israel organizations had already promised to contribute large sums to her Senate campaign.

Exactly one week before Giuliani bowed out of the race, Hillary asked Yasser Arafat's friend Hani Masri to host a secret fund-raiser for her at his Washington home. The Masri family poured hundreds of thousands of dollars into Democratic Party coffers just as the Clinton administration was pondering a $60 million government loan to Masri's Capital Investment Management Corporation. The two-hour-long reception at the Masri mansion netted a tidy $50,000 for Hillary's campaign.

The following month, Hillary showed up at another event to collect $50,000 from the American Muslim Alliance, an organization that advocates the use of force against the state of Israel. Later, Hillary claimed to have been unaware that the alliance had sponsored the fund-raiser. But then a thank-you letter written on White House stationery and bearing Hillary's signature turned up, as did a photograph of a smiling Hillary posing with a plaque the organization had bestowed on her. Things only got worse when it was revealed that a donation from another questionable group, the American Muslim Council, was disguised in the campaign's records as a contribution from the American "Museum" Council. "A typo," Hillary commented with a shrug. The only typo, it turned out, in the campaign's extensive list of donors submitted to the Federal Election Commission.

That June Hillary was, in the words of a state party official, "thrown into a panic" when crowds booed her as she marched up Fifth Avenue in New York's annual Israel Day Parade. Later, at a "Solidarity for Israel" rally in front of the Israeli consulate in Manhattan, she was booed off the stage.

Polls were showing Hillary and "Little Ricky," as she now referred to Lazio, as dead even. Convinced the race would be decided by a hairbreadth, she was more determined than ever to do whatever she could to win over Jewish voters. Not long after the "Solidarity for Israel" debacle, Hillary's handlers briefed her on the situation in New Square, a tiny Hasidic enclave about thirty miles northwest of Manhattan. There, four prominent members of the Orthodox Jewish Skver sect had been convicted in 1999 of bilking government aid programs to the tune of $30 million and funneling the cash back into New Square's yeshiva.

New Square had voted as a bloc in previous elections, and Hillary's advisers suggested she make a campaign stop there. She was also warned that four of the community's leaders were serving time in a federal prison for stealing millions of taxpayer dollars, and that the rabbis had been lobbying aggressively to have the four men's sentences commuted. Giuliani refused to help, as did Lazio.

A campaign staffer cautioned Hillary that the issue of a presidential pardon might come up. "So?" the candidate asked incredulously.

On August 8, 2000, Hillary made her pilgrimage to New Square, following Hasidic custom by covering her head with a scarf before sitting down to talk with village leaders. Positioned across a coffee table from the rabbis and talking through a tall flower arrangement that served as the Hasidims' traditional screen between the sexes, Hillary calmly spelled out what she was willing to do to improve New Square's health care services. The subject of pardons, both Hillary and her hosts would later insist, was never raised.

Six days later, on the opening night of the Democratic convention in Los Angeles, Hillary thanked the party faithful "for your support and faith in good times—and in bad." Surrounded by Democratic women senators Dianne Feinstein, Barbara Boxer, Blanche Lincoln, Mary Landrieu, Barbara Mikulski, and Patty Murray, Hillary seized the opportunity to position herself as a figure of national importance in her own right. "Bill and I are closing one chapter of our lives," she declared, "and soon we'll be starting a new one." When Bill followed her onstage to chants of "four more years," the Clintons smiled and subconsciously nodded in agreement. If all went according to The Plan . . .

The rabbis of New Square, meanwhile, would have to wait for several months before Hillary sent them a signal. The day before the election, a letter from the President was posted in the entrance hall of New Square's main synagogue. In it, Bill Clinton said he was looking forward to visiting the village in the near future. Once the votes were tallied the next day, New Square delivered Hillary the biggest victory margin of any community in the state—1,359 votes to only 10 votes for her opponent, Rick Lazio. In contrast, the two neighboring Hasidic villages, where the issue of a presidential pardon played no role in the election, voted 3,480 to 152 *against* Hillary.

"We had very strong support in the Jewish community, particularly in the towns surrounding New Square," Lazio said. "But from the very beginning, we were told not to come to New Square itself. They'd say, 'Don't bother to send anyone, there's nothing you can do. . . . We're terribly sorry.' We didn't know what it was at the time. Of course, we found out eventually."

Three days before Christmas 2000, Grand Rabbi David Twersky and several other New Square leaders would meet with Bill and Hillary in the White House Map Room. Sobbing, the grand rabbi begged the President to pardon Kalman Stern, Jacob Elbaum, Benjamin Berger, and David Goldstein.

Hillary would later tell federal investigators that she did not express an opinion at the meeting, and that at no time did she ask her husband to intervene. She knew, of course, that she didn't have to. "I don't believe that, before the election, anyone was dumb enough to say, 'Hey, we'll trade this for this,'" observed another New Square rabbi, Ronnie Greenwald. "But post-election, I think they took advantage of a quid pro quo and said, 'We helped you, we went all-out for you. If you can help us . . .'" Not surprisingly, the New Square embezzlers would get their commutations from the grateful candidate's husband.

Hillary would later express anger at Al Gore for distancing himself from Bill Clinton and the Clintons' much-debated legacy during the presidential race against George W. Bush. But no one understood better than Hillary that many lingering resentments over Bill's behavior could cost her crucial swing votes upstate and in the suburbs. Determined to disentangle her political career from her husband's, Hillary kept Bill's appearances on her behalf to a minimum.

Yet the President did play a key backstage role. While he did not attend most staff meetings, he worked the phones on her behalf, tinkered with her speeches, and—most important—used the power and prestige of his office to make the hard deals that would reap dollars and votes for his wife.

She did not want Bill around—he made her too nervous—as she prepared for the first of three debates with Lazio. Not that she needed his help. During the first debate, Lazio left his podium to invade Hillary's personal space, waving papers in her face and demanding that she sign an agreement forswearing soft-money contributions. She refused at the time, but that did not resonate with voters. What did make a lasting—and decidedly negative—impression was Lazio's perceived bullying of the First Lady.

"That's how it was spun," said Lazio, who claimed that at the time Hillary ("She's a bare-knuckles politician") looked anything but frightened when he stepped up to her. "Hillary *despises* being challenged. I remember looking at her and you could see she was thinking, 'I can't believe this guy is challenging what I have to say.'" Lazio went on to dismiss the notion, put forward by Hillary's camp in the days following the debate, that she somehow felt physically threatened when he stepped up to her during the debate. "It's the most ludicrous idea," Lazio said, "that she was this fragile flower, very vulnerable and passive. They really felt it was pure effrontery to contradict this woman who is very aggressive, very tough, and more than willing to throw a punch." What of the newer, gentler Hillary? "People don't change," Lazio said. "You are who you are. You don't become a patient, kind, ethical person if you weren't that beforehand."

Lazio was unable to overcome his late start in the campaign—not to mention Hillary's star power. "She'd fly into these little upstate towns aboard an Air Force plane, with this Secret Service entourage, and out steps the reigning First Lady," Lazio said. "It was pure Hollywood."

On election night, Hillary knew she would win, but not by how much. While her supporters waited for her in the ballroom of New York's Grand Hyatt Hotel, Hillary, swathed in a terry-cloth robe, was being tended to by her hairdresser and makeup artist when Chelsea burst in with the news. Hillary had won by a staggering 12 percent, 55 percent to Lazio's 43 percent. In the end, the victory had less to do with issues than it did with name recognition. "New Yorkers want someone bigger than life," noted *Time* magazine's Margaret Carlson, "and Little Ricky was no match for a vanity candidate like Hillary."

While mobile confetti guns showered the crowd with red, white, and blue confetti, Hillary thanked the multitude. "We started this great effort on a sunny July morning in Pindars Corner on Pat and Liz Moynihan's beautiful farm," she began, "and 62 counties, 16 months, three debates, two opponents, and six black pantsuits later, because of you we are there." She went on, in typically ponderous and penitential Hillary Clinton fashion, to remind her supporters of all the hard work that lay ahead.

Having carefully stage-managed her own celebration, Hillary stood between Chelsea and her Senate colleague Chuck Schumer, with Bill pushed off to one side. Red-faced and puffy-eyed, the President dabbed at tears of pride. Hillary was all business. "After eight years with a title and no portfolio," she said triumphantly, "I was now 'Senator-elect.' "

She would have to be, if Republican Majority Leader Trent Lott's reaction was any indication. "I tell you one thing," Lott said, "when this Hillary gets to the Senate, if she does—maybe lightning will strike and she won't—she will be one of one hundred, and we won't let her forget it."

Hillary's opponent in the Senate race had no doubt that she intended to pursue higher office. "I don't think she ran simply to grow old in a Senate seat," Rick Lazio said. "She's somebody who enjoys power, and when you've tasted the sort of absolute power she's accustomed to . . ."

Hillary's victory was upstaged by the seesawing results of the presidential race, closest in the nation's history. For the next "36 days from hell," as Laura Bush's staffers would call it, the Gore and Bush legal teams battled it out all the way to the U.S. Supreme Court. At stake were Florida's twenty-five electoral votes, and the presidency.

Hillary was enraged when, despite the fact that Al Gore had

won the popular vote by more than a half-million ballots, the Supreme Court voted 5–4 to stop recounting votes in Florida. Slamming the decision as "indefensible" and a "blatant abuse of judicial power," she pledged to back a constitutional amendment that would eliminate the electoral college altogether. Few of her colleagues, however, would have the stomach for such drastic measures. By the time George W. Bush was sworn into office, Hillary had stricken the issue from her repertoire.

Yet even before she was actually sworn in, Hillary began pulling the levers of power, calling on her old friends in the Senate to help her land spots on key committees. Hillary wisely courted Robert Byrd, the man who had repeatedly denounced the Clintons for their "egregious arrogance," and was rewarded with a seat on the Budget Committee. Nevada Senator Harry Reid offered her a spot on the Environment and Public Works Committees, and Ted Kennedy made room for her on his HELP (Health, Education, Labor and Pensions) Committee. Later, Hillary would also land a seat on the Armed Services Committee.

Just as she pulled strings to land the kind of committee assignments that would never go to a freshman senator, Hillary later lobbied for a Capitol Hill office befitting her stature. "She's not just another senator, no matter what they say," noted one awestruck Capitol Hill intern. "She's American royalty, more like a queen." After a six-week stint in the basement of the Dirksen Senate Office Building—the customary habitat of most newly elected senators—Hillary, who ranked ninety-eighth in seniority, was nevertheless catapulted to the majestic top-floor office suite previously occupied by Senator Moynihan in the Russell Building. On the walls hung portraits of New York senators who preceded her, ranging from Aaron Burr and Martin Van Buren to Bobby Kennedy. If its soaring ceilings and pink marble fireplace weren't indication enough of Hillary's stature, one only had to look at the man who occupied the similarly grand office next door: then–Senate Majority Leader Trent Lott.

Similarly, Hillary would insist that her Manhattan offices one block from the Waldorf-Astoria not only be more than twice as large as her fellow New York senator's—7,900 square feet to Chuck Schumer's 3,900 square feet—but that they cost the taxpayer more than twice as much. While Schumer paid $209,532 a year for his offices, the rent on Hillary's suite was $514,148 annually—far and away the most spent by any U.S. senator on office space.

Yet Hillary's Manhattan landgrab paled in comparison to her husband's. When Bill announced that he would be spending $800,000 a year to rent the entire fifty-sixth floor of Carnegie Hall Towers on Central Park South—three times what had been spent on the offices of any other ex-President—a public outcry ensued. Eventually, Clinton backed down, choosing instead to move into offices at 55 West 125th Street in the heart of Harlem. Nevertheless, his offices would—at a yearly rent of $300,000—still cost the taxpayer more than those of any of his peers.

Since her aspirations extended beyond the Senate, Hillary also had to maintain a residence in Washington that made it possible for her to host large groups of powerful people. By Christmas, she had put a down payment on a "Whitehaven," a $2.7 million, gabled, three-story brick Georgian manse with circular drive just off Washington's Embassy Row on Whitehaven Street. The spacious rooms, decorated in shades of yellow and pink by the same interior design team that did the nearby British embassy, would soon be filled with power brokers of every conceivable stripe. "She is always warm and gracious," said one frequent guest, "and she has a self-deprecating humor, and that hearty laugh—like a man's—can be startling." But he also likened his meetings with Hillary, all of which had a very specific purpose, to "calling on the Godfather."

On January 3, 2001, Bill, Chelsea, and Dorothy Rodham watched from the visitors' gallery as Hillary took the oath of office on the Senate floor. "In a classic Washington tableau of power-worship, hypocrisy and redemption," *New York Times* reporter Allison Mitchell observed, "the Senate took on the look of a receiving line, as a parade of senators came to welcome Mrs. Clinton, air-kissing, back-patting, and handshaking. . . ." The pièce de résistance came courtesy of ninety-eight-year-old South Carolina Republican Strom Thurman, who unnerved the First Lady senator with a disquietingly ardent embrace. This display of obeisance, taking place just out of TV camera range, left little doubt as to who was now the most powerful force in the Democratic Party. "There was a whole lot of ring-kissing going on," said one observer. MSNBC's Chris Matthews agreed that Hillary was "the number one Democratic senator right now. She's the boss. Hillary is calling the shots."

Of course, no one actually believed Trent Lott's assertion that she was just one of one hundred senators. Gallop polls showed that Hillary easily beat out Oprah as America's most admired woman—a position she would still be holding in 2004. Then there were the Secret Service agents who trailed Hillary—and would always trail her—wherever she went, creating consternation and envy among her status-conscious peers.

Still, Hillary was careful to start off her Senate career by flattering, cajoling, flirting with, and essentially catering to her fellow legislators' fragile male egos. "My gut is she'll handle it pretty well," said Connecticut Democrat Christopher Dodd. "She's smart. Smart is understanding that people are waiting here for any false move to jump all over her."

As far as the swearing-in ceremony was concerned, once was not enough for Hillary. She would reenact the solemn moment twice more for her supporters. The final reenactment would take place several days later at Madison Square Garden, where union executives and state party leaders watched as Al Gore administered the oath of office. Gore, who received several standing ovations, congratulated Hillary "not only on winning, not only on all the votes you received, but on getting your votes counted. . . ."

It was a strained moment for the Gores and the Clintons. Not long before, the two men had had a famous confrontation behind closed doors. Gore blamed his defeat on the outgoing President and the stink of scandal that trailed in the Clintons' wake. Clinton slammed Gore for betraying the Clinton legacy by not running on the administration's record. The acrimony would spill over to their wives. Hillary had always dismissed Tipper as an intellectual lightweight, and Tipper regarded Hillary as, in the words of an aide, "grasping." After Gore's defeat, the two women who had barnstormed the country on behalf of their husbands in 1992 and 1996 refused to speak to each other.

Once back in Washington, Hillary reminded Bill that this would be the last time he would be able to issue presidential pardons—a power that, they had both discovered, could be used to their political advantage. At Hillary's urging, he cast about for criminals worthy of executive clemency.

On January 15, the day the Clintons closed on the purchase of their Embassy Row house, Hillary was ebullient but her husband was plainly exhausted. There was no time for him to relax and enjoy his last few days in office, he told one of the lawyers present at the house closing, because he had to plow through a mountain of requests—nearly five hundred in all—for commutations and pardons. A few days later, returning from a quick trip to Arkansas aboard *Air Force One,* the President strolled into the plane's press section and drawled, "You got anybody you want to pardon?"

The Clintons' eagerness to pardon, though politically motivated, was also fueled in part by their hatred of special prosecutors in general and Ken Starr in particular. Hillary, who saw herself and her husband as the victims of Starr and his overzealous minions, said she was now more sensitive than ever to the fact that "lots of people are wrongly persecuted."

During the Clintons' last forty-eight hours in office, the White House ordered the Justice Department and the FBI to do background checks on scores of possible candidates. There was only one exception: Bill's younger brother, Roger, who had been convicted of trafficking in cocaine in 1985. Bill was well aware that the problematic Roger (Secret Service code name: "Headache") was the target of several FBI inquiries into alleged influence peddling. He was also highly resentful of the FBI's dogged pursuit of investigations into a number of Clinton scandals. The President wanted Headache's pardon handled by more Clinton-friendly higher-ups in the Justice Department.

Roger wanted more than just a pardon for himself; he was also actively lobbying to have his brother pardon six of his buddies. But Roger was not the only family member deeply involved in the process. Hillary's Brobdingnagian younger brother Hugh Rodham, now a Florida lawyer, was pushing hard to have the President pardon drug kingpin Carlos Vignali and Almon Glenn Braswell, an herbal-remedy magnate convicted of perjury and mail fraud.

Bill and Hillary knew that Braswell was worth millions, and that Vignali's family had contributed heavily to the Democratic Party since Carlos's imprisonment in 1994. But Hillary would later claim she had no idea that "Hughie," who had stayed at the White House so frequently that he now required more packing crates than Chelsea, stood to collect $400,000 in fees for securing the pardons.

Hillary would later deny it, but in those last few days in the White House, she conferred several times with her husband about who would and would not appear on the list of pardons he was preparing. "Hillary was involved from the get-go in making sure that several pardons went through," insisted a longtime Arkansas friend with an intimate knowledge of at least four of the clemency cases from the President's home state. "Hillary brought him names, pushed them along, and followed up with phone calls. In other cases, he ran names by her because he didn't want anything he did to backfire and cause her grief. But obviously they both figured that after the whole Bush–Gore election battle . . . the pardons would just slide by and nobody would notice."

At least two were foregone conclusions: Roger, and the Clintons' ex–Whitewater partner Susan McDougal, who served two years in prison—first for fraud in the land deal, and then for refusing to testify against Bill. Apparently unperturbed by Susan's alleged affair with her husband, Hillary would go on to salute McDougal's loyalty in her autobiography.

Nearly all who made the final cut were recommended by family, friends, or fund-raisers. Longtime Clinton pal Jesse Jackson, who had just tearfully confessed that he had fathered a child out of wedlock even as he counseled the Clinton family during the Monica Lewinksy mess, successfully sought clemency for two aides and former congressman Mel Reynolds, sentenced to six and a half years for wire and bank fraud (after serving a two-year term for having sex with a teenager). Harry Thomason helped win pardons for two Arkansas tax evaders, and Democratic National Committee Chairman Terry McAuliffe convinced Clinton to pardon lobbyist James Lake, convicted of masterminding an unlawful campaign contribution scheme.

Clinton advisers William Kennedy III, David Dreyer, and James Hamilton asked for and got clemency for various cocaine traffickers, tax cheats, and money launderers. For obvious reasons, Bill sympathized with his former Housing and Urban Affairs Secretary Henry Cisneros, who made secret payments to his mistress and then lied to federal agents about it; Clinton pardoned Cisneros *and* his mistress.

Portly New York Congressman Jerrold Nadler, one of Bill Clinton's most vocal defenders during the impeachment proceedings and a major booster of Hillary's Senate candidacy, interceded on behalf of 1960s radical Susan Rosenberg. A member of the violent Weather Underground, Rosenberg was arrested in New Jersey in 1984 with 740 pounds of dynamite and a machine gun in her car. She was also a suspect in the 1981 robbery by Weather Underground radicals of an armored car that left a Brinks guard and two police officers dead. Overriding Justice Department objections, Clinton pardoned Rosenberg just as he had pardoned the FALN terrorists the year before.

Another pardon application was of particular importance to New York's newest senator. Wealthy Manhattan socialite-songwriter Denise Rich had written to the President pleading for him to pardon her ex-husband, notorious white-collar fugitive Marc Rich. The Belgian-born billionaire, charged in 1983 with evading $48 million in taxes and illegally trading with Iran during the hostage crisis, had fled to Switzerland with Denise and his business partner, Pincus Green.

For years the glamorous, well-connected Denise had ranked as one of the Democrats' most generous donors, contributing more than $1.5 million out of her own pocket to the party, another $450,000 to the Clinton Presidential Library in Little Rock, and $70,000 to Hillary's campaign. She was best known for the fundraisers she threw in her lavish, twenty-five-thousand-square-foot penthouse apartment on Fifth Avenue.

Not surprisingly, Denise quickly became an intimate of both Clintons. "They hug and are very close," another friend, actress Jane Seymour, said of the relationship between Denise and the First Couple.

Beth Dozoretz, former finance chief of the Democratic National Committee, was also backing Rich's pardon. Rich had taken the added step of hiring former Clinton White House counsel Jack Quinn as his lawyer, but Denise and Beth Dozoretz—

who, according to Secret Service logs, racked up ninety-five White House visits between them—were, as Marc Rich boasted to a friend, "the ones who have really got Hillary's ear."

The President would wait until the morning of this last day in office and then, over strong objections from the Justice Department and all of his own top aides, add Rich's name to the list. "He did it," one of Bill's closest advisers would later explain, "for Denise." And, by virtue of the money she would continue to funnel Hillary's way, for New York's freshman senator.

In coming weeks, the scandal over "Pardongate" would assume a life all its own. At first, Hillary would do what she had always done: claim total innocence. She had nothing to do with the pardons, she insisted. Nothing at all. As federal investigators closed in, however, she was forced to concede that she "may have" played a part in the selection process. "When it became apparent around Christmas that people knew that the President was considering pardons, there were many people who spoke to me, or you know, asked me to pass on information. You know, people would hand me envelopes, I would just pass them. . . ."

Weeks later, Hillary would be sitting in a Washington theater watching *Crouching Tiger, Hidden Dragon* when an aide called on the senator's cell phone to tell her about Hughie's $400,000 payday. Publicly, Hillary would only admit to being "very disappointed . . . and deeply saddened" by Hughie's lucrative deal. She called for her brother to return the money.

Privately, Hillary raged at Hughie Rodham, who was staying with the First Family during those final days in the Executive Mansion and yet had said nothing to his sister about the payoff. "How could you," she demanded to know, "be so goddamned stupid?"

Democratic Party leaders would soon be asking the Clintons the same thing. Even the Clintons' staunchest allies conceded that Bill and Hillary had gone too far this time. *New York Times* editorial writers branded the pardons as nothing less than an "inexcusable abuse" of executive power. Jimmy Carter broke the code of the presidential brotherhood by calling Clinton's pardon of Rich "disgraceful." Hillary's fellow New York senator, Chuck Schumer, also condemned the Marc Rich pardon.

Even as Denise Rich and Beth Dozoretz pled the Fifth Amendment at hearings before the House Government Reform Committee, Terry McAuliffe admitted the Rich pardon was simply "wrong." And West Virginia's courtly Robert Byrd, Hillary's mentor in the Senate, would claim to be "disgusted" by the President's last official actions in office. Byrd characterized them as "malodorous."

In their final days at the White House, however, the Clintons had not the slightest inkling of the controversy that awaited them. Instead, Hillary devoted a considerable amount of time to mapping out the family's financial future.

Hillary now set her sights on, as she told friends, "finally having some nice things we can call our own." Even before giving up her lucrative Little Rock law practice to become First Lady, Hillary had frequently complained of the sacrifices she had had to make so that Bill could pursue his presidential ambitions. It had been years since Dick Morris warned Hillary that her plan to put in a swimming pool at the Governor's Mansion in Little Rock might be resented by voters in a state as impoverished as Arkansas. "Why," she shouted at Morris in reply, "can't we live like normal people?"

Hillary was well on her way to living well when, in mid-December, she began auditioning publishers in the White House Diplomacy Room. It was vital that she ink a deal before January 3, 2001, when, upon being sworn into office, she would have been required to run it by the Senate Ethics Committee.

Hillary later justified this apparent scramble for a stratospheric book deal by pointing out that the First Couple faced more than $4 million in legal fees stemming from the various investigations into their conduct. But she also knew that several defense funds were already well on the way to covering the entire amount with hefty contributions from wealthy FOBs and FOHs.

While industry insiders speculated that she would be offered a $5 million advance for her memoirs, she ended up accepting Simon & Schuster's $8 million—just $500,000 shy of Pope John Paul II's record-shattering advance. Even after signing on the dotted line, Hillary was worried that Senate watchdogs might do something to prevent her from collecting all the money she was due. "She was pushing hard," recalled someone close to the negotiations, "to get the whole advance up front."

Few remembered that, just six years earlier, Hillary and Bill had both denounced Newt Gingrich's $4.5 million book deal, even though at the time there were no limits on what a congressman might accept in the way of payment. Nevertheless, a chastened Gingrich gave back the $4.5 million and agreed to accept an advance of only $1. But when a staffer reluctantly showed Hillary an editorial in the *New York Times* condemning her $8 million contract, she merely shrugged. "Screw 'em," the First Lady said yet again, then returned to her paperwork.

Hillary also made certain that Bill would be doing his part. "Bill was happy to go out on the lecture circuit and make some big money," a former staffer remarked. "But he wanted to take it slow at first and be selective about which offers he accepted. Hillary said, 'No way, buddy, you're doing it all.' " Weeks before the Clintons exited the White House, speaking engagements were being booked for the soon-to-be ex-President at fees ranging from $100,000 to $450,000. One postpresidential payday Bill seriously considered was vetoed by Hillary as "just too tacky": $2 million to appear in a Super Bowl commercial.

(Bill was already well aware of his earning potential and determined to exploit it to the fullest. Although he was vocal in his public opposition to George W. Bush's planned tax cuts, Clinton applauded them in private. Over dinner one evening with Oxford Professor Alan Ryan, Clinton said Bush's cuts would be "good for people who are as rich as I'm about to be." Said Ryan: "He made it quite clear he expects to make a colossal amount of money very fast." As it turned out, Clinton would earn $9.2 million in speaking fees alone his first year out of office—and end up breaking the pope's record by signing a $12 million book deal with Alfred A. Knopf.)

After eighteen years in public housing—first at the Arkansas Governor's Mansion and then at the White House—Hillary was finally prepared for the transition to the Embassy Row and Chappaqua houses. Over the previous year, she had relied on her old Arkansas-based interior decorator pal Kaki Hockersmith to help transform the relatively modest Chappaqua house into a residence befitting a United States senator and her spouse, the ex-President.

Hillary had, after all, become accustomed to a certain lifestyle. At one point that final week, she was escorting a guest through the family residence when she stopped to admire a gigantic spray of yellow roses and chrysanthemums in the East Sitting Hall. "This is one of the great things about living in the White House," she said with a sigh. Moving on down the cavernous Center Hall, they passed the Cézanne and then the de Kooning before stopping in front of Mary Cassatt's painting of a woman and two children. "When we go to our house in New York, we take out things and I think, 'Doesn't that look nice?' And then I come back here," she said, "and there's the Mary Cassatt."

Hillary would later write that, during these last few days in office, she took the time to take a stroll through the White House Children's Garden with Chelsea, and to wander "from room to room making mental snapshots of all my favorite things."

As it turned out, she didn't plan on leaving all that much behind. At one point, Hillary walked the halls of the family residence with a clipboard-toting aide, pointing to the items she intended to take with her. In the solarium, Hillary plopped down on the $6,000 English-style sofa that had been a gift to the White House from Manhattan businessman Steve Mittman. "We're taking this," she told the aide, who hastily jotted down a description of the piece on a yellow legal pad. "And those," Hillary said, nodding toward four oversize club chairs—each valued at $2,843—that had also been a gift from Mittman.

As they moved on to the Yellow Oval Room, here were the two Henredon sofas worth $3,000 each, and the rattan chairs in the solarium, and the Aubusson rug in the First Lady's Sitting Room, and the kitchen table, oh, and of course the $7,375 worth of designer tables and chairs from our dear friend Denise Rich . . .

"On behalf of the Clinton Family," Hillary had written to Mittman and the others when they first donated the furnishings, "I want to express my sincere gratitude for your generous contribution to the White House." But now that the Clintons were departing, Hillary preferred to look at these pieces of furniture—$28,000 worth in all—as her personal property.

Not that this accounted for more than a small fraction of the items needed to furnish her two sizable residences. Hillary waxed nostalgic to *Ladies' Home Journal* reporter Meredith Berkman about what it was like unpacking old family treasures. "I don't even know where to start," Hillary gushed. "Old rocking chairs, old tables, old clocks, just everything . . . a lot of old pictures, knickknacks, memorabilia."

Away from reporters, she lamented the fact that, after nearly two decades of living in taxpayer-funded splendor, "all we've got to call our own is some old junk." Always adept at finding an angle—tax records show the Clintons had actually taken a deduction for donating used underwear to charity—Hillary was searching for a way to finance the furnishing of her new homes without drawing fire from the press.

After brainstorming with decorator Hockersmith, Hillary came up with a novel solution. Rather than asking for cash contributions outright, why not quietly register for gifts at a department store. "Like a bride!" Hillary said, cornflower blue eyes widening. Then she cut loose with her trademark, window-rattling laugh.

Hillary went ahead with her plan and registered with a suitably low-profile store, Borsheim's, in Omaha. On a fund-raising trip to Nebraska that December, she ducked into the store to peruse the merchandise. Again, she wanted to take advantage of the small window of time between her election to the Senate and her swearing-in, when she would be barred from accepting any gift valued at more than fifty dollars.

On January 19, 2001, Hillary watched from an upstairs window as two twenty-six-foot-long vans pulled up to the White House. Pushing dollies through the broad corridors of the second-floor residence, movers hauled away all the items Hillary had tagged—which now also included three television sets, a DVD player, a china cabinet from insurance tycoon Walter Kaye (who happened to be the man who first delivered Monica Lewinsky to the Oval Office), two Dale Chihuly glass sculptures worth $60,000, two Lenox bowls valued at over $50,000, $22,000 worth of china (Spode), $18,000 in silverware (Fabergé and Stafford), as well as individual decorative objects fashioned by Tiffany, Cartier, and Waterford.

Items worth $250 or less did not require public disclosure, so it appeared that Hillary undervalued many of them with that figure in mind. An Yves St. Laurent men's suit, for example, was listed as being worth $249—perhaps a tenth of its true cost. Hillary signed off on the final tally, and formally declared that the Clintons were taking more than $190,027 worth of furniture and "gifts" with them—compared to the $52,000 worth of furnishings George and Barbara Bush took with them. Eventually, government investigators would determine that Hillary and Bill were carting away merchandise valued at more than twice what they officially declared—in excess of *$400,000* worth.

"It was like watching Imelda Marcos stuff furs and jewelry into a pillowcase as she was getting ready to flee the presidential palace," said one veteran White House steward. "A lot of us were very upset by what we were seeing."

So, it seemed, were congressional investigators. Under pressure, Hillary finally agreed to return $50,000 worth of gifts originally intended for the White House, and later anted up an additional $86,000 to cover the value of items given to them during their last few months in office.

The flap over Hillary's Imelda Marcos impersonation paled in comparison to Pardongate. HIL MUM ON PARDONS, screamed the front-page headline in the New York *Daily News,* while the rival *New York Post* blared BEG PARDON, BUT HILLARY IS LYING LOW. *Time* marveled at how "breathtaking" it was to watch "a shiny new presidency" buried in a "freak mudslide. The debris hurtled by so fast that the *New York Times* editorial page seemed to run out of synonyms for disgust, revulsion and abuse . . . there seemed to be no end to the bodies that might float down the swollen river." Speaking for most of his colleagues on both sides of the aisle, Connecticut Congressman Christopher Shays characterized the pardon flap as "a really slimy affair. The more we look into it, the slimier we feel."

Hillary hunkered down and rode out the storm. Whenever she was asked about the pardons, she insisted she had nothing to do with them—that her husband was solely responsible for each and every one. Each new denial, couched in terms that made her sound like a victim, sounded much like the last: "I'm very regretful that it occurred, because I might have been able to prevent this from happening." "I don't know anything other than what has now come out." "You know, I did not have any involvement in the pardons. . . ." "I'm very disappointed. I'm very saddened, and I was very disturbed when I heard about it." "When I found out about my brother . . . I was heartbroken and shocked." "I knew nothing. . . ."

Yet there was no question that Hillary, who had been consulted by the President on everything from health care and education reform to Middle East policy and the bombing of Bosnia and Iraq, was, as one close Clinton associate put it, "deeply involved" in deciding who would and would not make the cut. "Are you kidding?" said another. "*Everybody* in the White House was talking pardons in those last few days. I mean, Roger Clinton and Hugh Rodham and Terry McAuliffe and Jesse Jackson and the Thomasons and about a hundred other folks were chatting up the Clintons about who was going to get pardoned—not to mention all the lawyers who were running around checking people out, and Hillary was the only one who didn't know about it?"

As the Clintons' late Commerce Secretary Ron Brown once said in response to a friend's question, "Is she in the loop? She *is* the loop."

Eventually, newly appointed Attorney General John Ashcroft, while condemning the pardons, would claim that he was powerless to do anything about them. The President's clemency powers were absolute. Ashcroft did ask for reforms in the way pardons were processed in the future, however, "so a travesty like this doesn't happen again."

While her husband wasted no time hitting the lecture circuit, Hillary burrowed into her work as a senator. She insisted that she was not reeling from the uproar over gifts and pardons. "I haven't felt distracted from my job," she said. But her colleagues disagreed. "You can see it on her face," said one. "This has been very difficult for her."

Bill was apparently having a better time of it. Dining with former Democratic Senator Bob Kerrey, former White House lawyer

Cheryl Mills, and others at trendy Babbo in Greenwich Village, Clinton roared his way through the retelling of an off-color joke involving former California Governor Jerry Brown and two lesbians. A C-SPAN microphone had picked up the joke the first time Kerrey told it to Clinton at a Bedford, New Hampshire, event in 1992, effectively sinking Kerrey's presidential campaign. This time, the former President was laughing so loud that Mills reportedly had to ask him to quiet down.

Hillary, meantime, busied herself with committee hearings, meeting with constituents, and the mountain of paperwork that came with being the cosponsor of 163 bills. "Many of my colleagues realize that sometimes they can get more attention," she said, "if I'm involved." More attention—and lots more cash. No sooner did she arrive on Capitol Hill than Hillary set up her very own Political Action Committee—HILLPAC—for the purpose of raising funds for Democratic candidates. Establishing a PAC of one's own was a highly unusual move for a freshman senator. More often than not, it was seen as a first step toward a presidential run. While denying that she had any intention of seeking the presidency in 2004, Hillary nonetheless held on to most of the money HILLPAC took in.

Indeed, Hillary's growing war chest only served to fuel speculation that she was contemplating a run for the White House. On April 5, 2001, after delivering a brief speech to the American Society of Newspaper Editors, Hillary again told reporters that entering the race for president was "not something I'm going to be doing."

"So, Senator Clinton," a reporter for the *New York Post* asked as she rushed toward her waiting car, "are you ruling out a run for president not just in 2004, but in 2008 and beyond?"

"Yes," Hillary replied.

The answer stunned her aides, and when the *Post* ran its headline on the next day's front page—HILL NO! CLINTON SAYS SHE'LL

NEVER RUN FOR PREZ—there was, predictably, chaos in Hillaryland. Clinton's aides urged her to issue a statement claiming she had been misunderstood, or that perhaps she had not heard the reporter's question clearly. But Hillary, who was married to the master of the precisely parsed nondenial denial, came up with her own "clarification." When asked the next day if she was really ruling out ever running for President, she answered, "I'm saying what I always said." What she always said was, simply, that she intended to serve out her first full term as senator, which was to expire in 2006. Over the next several days, she would respond to these persistent queries with the same exact, carefully weighed words—"I'm saying what I always said."

Republican strategist Nelson Warfield was among the many who remained resolutely skeptical. "I just think this is the latest chapter in the Hillary Clinton saga," he said. "I'm certain this does not end either her cogitation or the press' speculation."

For the most part, Hillary tried to keep her head low as she adjusted to life in the Senate. At times, however, she was taken to task by her colleague Chuck Schumer for trying to upstage him. For his part, New York's senior senator was no slouch when it came to the photo op; Schumer was so deft at self-promotion that other politicians who had been upstaged by him complained that they had been "schumed." On several occasions when Hillary went ahead and unilaterally announced some piece of legislation benefiting their state, Schumer personally berated members of her staff. "Tell your boss," he snapped at one of Hillaryland's junior members, "that we had a deal to consult each other *before* making announcements. I'm sick and tired of her trying to hog the limelight."

The junior senator for New York apologized profusely, promising never to step on her colleague's toes again. But according to one state party official who knew both senators, "Hillary knows exactly what she's doing. There's only so much credit to go around, and she wants it all. Besides, she doesn't like him any more than he likes her." Fuming over yet another perceived slight, Schumer spoke of Hillary in scathing terms. "It's no secret," the official said, "that he thinks she's a bitch." Privately, Hillary dismissed publicity-craving Schumer as "The Prima Donna."

Whatever animosity existed between Hillary and Schumer, it paled in comparison to what she felt for George W. Bush. By all accounts, Senator Clinton harbored what amounted to a profound hatred of the man who now occupied the White House. In private, Hillary often referred to George W. Bush as "Junior" or "Shrub"—the moniker employed by Texas Democrats—and inveighed against him for having "stolen" the election.

Hillary, who had been feuding with her predecessor Barbara Bush for years, also had little use for her replacement. It didn't help that during the campaign W portrayed his wife Laura as the anti-Hillary. "She's not always trying to butt in, and you know, compete," he said coyly. "There's nothing worse in the political arena than spouses competing."

Barbara Bush also held up her school-librarian daughter-in-law as the antidote to eight years of scandal and controversy. She believed Laura would be different from Hillary because Laura "would rather make a positive impact on the country. And I'm not criticizing Mrs. Clinton. But it's like oil and water. . . . They're two different people. I think Laura thinks of others."

Hillary was accustomed to Barbara's catty remarks, and viewed W's beloved white-haired mother as a worthy adversary. The new First Lady was another matter. Hillary viewed Laura, said an aide in the New York Senate campaign, as "very nice, very dull, and not the brightest light on the porch."

One First Lady both women approved of was Jackie Kennedy. "She had the most marvelous taste," said Laura, who set out to restore the residence so that it was "just the way that Jackie left it." She promptly brought up Jackie's favorite velvet-upholstered chairs from the White House cellar and pulled down the heavy print draperies the Clintons had installed in the upstairs yellow living room. By way of erasing the heavy-handed influence of Hillary's Arkansas decorator Kaki Hockersmith ("Tacky Khaki" to some in the interior design business), Laura repainted the walls, moved furnishings and antiques from room to room, and brought hidden treasures out of storage. Most important, she moved the First Lady's offices back to the East Wing, where all Presidents' wives had had their offices prior to Hillary.

Senator Clinton had no intention of sharing the spotlight with Laura Bush at the April 2001 gala opening of *Jacqueline Kennedy: The White House Years,* a lavish fashion exhibit at New York's Metropolitan Museum. Caroline Kennedy, who presided over the event, was unaware that Laura had long admired her mother. Caroline virtually ignored Laura, thanking the new First Lady "for coming tonight" before moving on to other guests. Hillary, meantime, had studiously waited for Laura to leave before making her own entrance in a floor-length leopard-print taffeta gown by Oscar de la Renta. Heaping praise on Clinton, Caroline proclaimed her the woman who "interpreted the role of First Lady for our times."

Back in Washington, Hillary seethed as the Republicans systematically dismantled many Clinton-era policies she had helped put in place. She told the Democrats' leader in the Senate, Tom Daschle, that it was time to open fire on Bush. The first step, she said, was to set up a War Room like the one she and Bill used to obstruct, intimidate, discredit, discourage, and destroy their opponents.

Such strong-arm tactics were of little use on Capitol Hill, Daschle explained to the freshman senator. Diplomacy, backroom deals, and decorum were what got things done in the Senate. Hillary took Daschle's advice—sort of. While she worked amicably with her colleagues on both sides of the aisle to accomplish legislative goals, Hillary also seemed to delight in attacking every administration official who appeared before one of the Senate committees she served on. After watching her take apart Secretary of State Colin Powell, one Republican spectator observed, "You have to hand it to that Hillary. She may be wicked, but she's effective."

Hillary's favorite target, of course, was the President. After he had been in office less than three months, Hillary lashed out at W for his policies on the environment, education, and health care. "He's trying to turn back the clock fifty or sixty years," she declared, "not just the Clinton Administration, they want to turn the clock back on the Roosevelt Administration. . . . The President's been on a charm offensive, but his administration is on a *harm* offensive."

By this time, it appeared to Hillary that she might be able to put the pardon controversy to rest once and for all. Toward that end, she even cosponsored a bill with Pennsylvania Republican Arlen Specter aimed at tightening the rules on presidential pardons. Only a few months earlier, Specter had raised the possibility of impeaching Clinton again—even though he had already left office—over the pardon issue.

Scandal reared its ugly head in April, however, when Denise Rich was suddenly telling her side of the pardon story on television and in *People* and *Vanity Fair* magazines. Although Rich denied that she had had an affair with the President, she coyly characterized their relationship in interview after interview as "very special."

Hillary was annoyed that, once again, tongues were wagging about her husband and another woman. But not so bothered that she would risk alienating a key New York fund-raiser. "Hillary looks the other away," said a former Clinton staffer, "when big money is involved."

That was also true of other women—attractive women with deep, deep pockets—linked to Bill Clinton during this period. One of those mentioned in several published reports in the spring of 2001 was Patricia Duff, who numbered among her four ex-husbands Hollywood studio executive Michael Medavoy and billionaire cosmetics tycoon Ron Perelman. Duff became friends with Bill when she raised millions for his 1992 campaign, and was a frequent guest at the White House throughout the Clinton administration. Duff, who served as executive director of the Women's Leadership Forum, signed on as co-chair of fund-raising for the President's reelection campaign. She didn't have to go far. With her then-husband Perelman, whose worth was estimated at over $2 billion, Duff threw a fund-raising bash in Palm Beach that netted $1 million. Bill rewarded Duff with a presidential appointment to the Library of Congress Trust Fund Board.

While the beautiful Georgetown University graduate denied that there was ever any intimacy between them, or that there was ever even a time when she was alone with the President, rumors to the contrary persisted even after the Clintons exited the White House. It didn't help that, after returning with her husband from one of her many stays in the Lincoln Bedroom, Duff gushed to guests at a Hollywood dinner party that Bill was "a full-service president." When the remark leaked out, she hastily tried to explain that she was only complimenting Bill on the extent of his hospitality.

The Clintons were nothing if not experts at spin control, particularly when it came to their marriage. Three years earlier, they had been photographed dancing in their swimsuits on a St. Thomas beach—a shot, Hillary claimed, that had been taken "unbeknownst to us" by a French photographer on a public beach across the bay. Hillary was satisfied that her response—"What fifty-year-old woman wants to be photographed from behind in her bathing suit?"—had convinced everyone that the tender moment captured on film was indeed genuine.

Since, in Hillary's mind, this ploy seemed to have worked so well before, the Clintons jetted off in April 2001 on a "top-secret" vacation in the Dominican Republic with friends Oscar and Annette de la Renta and a few other top Democratic Party contributors. The Clintons' loving behavior would be duly reported by other guests at the exclusive Punta Cana resort, and as soon as she returned to Washington, Hillary burbled to the press about how romantic the getaway had been. "You know what was great about it?" she said, laughing. "There were no pictures of me in my bathing suit!" The Clintons would not actually be seen together in public until May—the first time since their stormy departure from the White House four months earlier.

However, it was only a matter of months before Hillary was once again betrayed by her husband, who apparently was very interested in seeing someone else in a bathing suit—a very skimpy bathing suit. Jetting to São Paulo, Brazil, for a $250,000 speaking engagement, Bill made a secret side trip to Rio de Janeiro on August 27, 2001. Once there, he and his Secret Service detail stopped off at the Blue Man beachwear boutique, where the former president spent $116 on two bikinis and three sarongs. Then, at 12:30 P.M., he checked into the $1,500-a-night Presidential Suite of the Copacabana Palace hotel. He left, hotel records would show, seven hours later.

For the time being, Hillary knew nothing of the episode. Instead, she was coping with an embarrassing incident closer to home. That same month, her brother Tony was caught having sex with another man's girlfriend and beaten to a pulp. Tony, recently separated from his wife Nicole, daughter of California Senator Barbara Boxer, was in the living room of the Rodham family's summer cottage in Lake Winola, Pennsylvania, when he was reportedly caught in flagrante delicto. His assailant was charged with assault, criminal trespass, and burglary. When told of the incident, Hillary rolled her eyes and sighed. "What next?" she asked.

That summer Hillary did some pummeling of her own, bashing President Bush as part of her plan to position herself as the Democrats' attack dog. Denise Rich was still in the middle of her publicity blitz when Hillary ripped into Bush on judges, school construction, school testing, air quality, work-protection rules, energy, taxes, and abortion. She mocked his missile defense ideas, accused him of increasing military spending at the expense of social programs, and cast the Senate's sole "no" vote on two of Bush's Justice Department appointees. Not coincidentally, both men had been lawyers on the Senate Whitewater Committee.

Dorothy Rodham, who was still bragging about how Hillary used to "beat up on" the boys in their Park Ridge neighborhood, was not at all surprised that her daughter embraced the role of political pit bull. With Chappaqua-based Bill on the lecture circuit and Stanford graduate Chelsea off to retrace her father's footsteps at Oxford, Hillary could devote virtually all of her time to building a national power base of her own. Although in private she sometimes referred to her fellow Democratic senators as "wimps," she was delighted to be the party's point person when it came to skewering Bush.

Hillary flailed away at the Republicans' alleged insensitivity toward working-class Americans, though on those rare occasions when she had to pay for something out of her own pocket, she seemed somewhat less than generous. After taking actress and longtime Democrat Lauren Bacall to lunch at New York's Russian Tea Room, Hillary picked up the $300 check. As they left, Bacall was reportedly stunned to see that Hillary had left only $15—5 percent—for the tip. As Hillary chatted with fans on the way out of the restaurant, Bacall made her way back to the waitress and slipped her $100. "The Senator," she said by way of an apology, "must have made a mistake."

On September 11, 2001, Bill was collecting a $150,000 speaking fee in Australia and Hillary was in Washington preparing yet another attack on George W. Bush. Hillary was on the phone shortly after 9 A.M. when an aide came running into her office and told her to turn on her television set. Only minutes before, an airliner had crashed into one of the Twin Towers of the World Trade Center. Like millions of other Americans, Hillary sat, mouth agape, watching as a second airliner slammed into the other tower. "Bin Laden did this," she said, then stiffened as it suddenly dawned on her just how close to home bin Laden had struck. "Oh my God," she murmured, her face turning ashen.

"What's wrong?" the aide asked.

"Chelsea. She's in New York. . . ."

Winning that election was liberation day for Hillary.

—Don Jones, Hillary's youth pastor and friend, on her Senate win

■

Does she plan to run for President? *Of course.*

—Ed Koch, former New York mayor

I want to run something.

6

She was not born in New York. She had never lived there. She did not attend school there, and none of her ancestors hailed from there. She had, in fact, never spent more than a few days there before deciding that she wanted to use one of New York's two Senate seats as a springboard to the presidency.

Now, as one of New York's highest elected officials, Hillary would be tested—completely on her own—in a way she had never been tested before. Her first thoughts, understandably, went to Chelsea. As she would later tell Katie Couric on NBC's *Dateline,* Chelsea "had gone on what she thought would be a great jog. She was going down to Battery Park, she was going to go around the towers. She was going to get a cup of coffee and—that's when the plane hit!"

Not quite. When she first learned of the terrorist attacks, Chelsea was actually staying at the Union Square apartment of her best friend, Nicole Davison—over thirty blocks from Ground Zero. Davison, who had already gone to work, called Chelsea to tell her about the first plane hitting the north tower. At that point, Chelsea later recalled, "I stared senselessly at the television," watching the second plane hit.

Alone and afraid, Chelsea tried to get through to her mother, but her phone was dead. Suddenly overcome by "panic," she bolted out of the apartment and started walking downtown—toward Ground Zero—in search of a working pay phone. As people, many covered in ash, streamed by her in the opposite direction, Chelsea became disoriented, confused. Suddenly there was a deafening roar, and someone told her that one of the towers had fallen. Chelsea would later say that at the moment she feared for her life. She wasn't sure where she was at the time, but based on the landmarks she described, friends told her she must have been a dozen or so blocks from Ground Zero.

Given the inconsistencies in Chelsea and Hillary's stories, there were those—including some New York Democrats—who doubted their veracity. The earliest news reports did not put Chelsea so perilously close to the World Trade Center. Members of Charles Schumer's staff even suggested that, the more the press kept reporting that Schumer's daughter Jessica was just five blocks away from Ground Zero at Stuyvesant High School, the closer Chelsea got.

Strangely, Chelsea would have no trouble reaching her father in Australia; it would take hours for her to finally touch base with Hillary, who by then was strategizing with staffers about how the junior senator from New York should respond to attacks on her constituents. How she handled the next few days would, Hillary quickly realized, have a profound impact on her future presidential prospects.

At the time of the attacks, President Bush was speaking to schoolchildren in Sarasota, Florida. Rather than return immediately to Washington, he followed the advice of those who feared for his safety and spent ten hours zigzagging from one air base to another aboard *Air Force One*. Predictably, a number of Democrats seized on this fact to criticize Bush, implying that the commander in chief had behaved in a less-than-heroic fashion.

Hillary was not one of them. Despite her reputation as one of President Bush's harshest and most outspoken critics, Senator Clinton now became one of his strongest supporters. Giving a speech on the floor of the Senate the day after the attacks, Hillary expressed her "strong support for the president not only as the senator from New York but as someone who for eight years has some sense of the burdens and responsibilities that fall on the shoulders of the human being we make our president. It is an awesome and oftentimes awful responsibility for any person."

She later told reporters that she was "100 percent behind Bush's handling of the terrorist attacks," and thanked the President for his pledge of federal help in the massive search-and-rescue effort that still lay ahead. "We will stand united behind our president," she said, "to make very clear that not only those who harbor terrorists but also those who give any aid or comfort whatsoever face the wrath of America. You are either with America in her time of need, or you are not."

Not all of Hillary's supporters were so willing to drop partisan differences for the sake of the country. One New Yorker who had contributed to Hillary's Senate campaign breathed a sigh of relief on 9/11 when, after frantically trying to reach his wife, he finally located her safely ensconced in her midtown Manhattan office. In the same conversation, Hillary's supporter was informed that conservative commentator Barbara Olson had been among those killed aboard the plane that crashed into the Pentagon. "Well," he snickered, "I can't say I'm sorry."

Two days after the attack, Bush called Mayor Giuliani and New York Governor George Pataki from the Oval Office to pledge his support. That same day, the President and his national security team met with Hillary and Schumer. Although she would later claim on many occasions to have fought hard for $20 billion in federal money for the city, Bush had made up his mind to back the request the minute he heard it. "Before we could say much of anything," Hillary later recalled of the meeting, "the President told us he would support the $20 billion in federal aide we had asked for."

Over the next several days, Bush and Giuliani—two of Hillary's political archenemies—rose to the task of pulling the nation together in a time of national crisis. Hillary praised Giuliani in particular for his courageous leadership during New York's darkest hour. At the same time, she fully appreciated the fact that 9/11 had made Giuliani, now being hailed as "America's Mayor," that much more formidable. "If she wanted to run for president in 2008, Giuliani could be the GOP nominee," observed a political analyst who had done work for both Clintons. "Or at the very least, he could go up against her when she runs for reelection. Anything that built up Rudy Giuliani's reputation—or for that matter Bush's—had to be a bad thing, as far as Hillary was concerned."

At times, Hillary had a difficult time concealing her true feelings about W. During the President's address to a joint session of Congress just days after the attacks, television cameras caught Hillary rolling her eyes and shaking her head. Her aides would later claim that the disapproving gesture had nothing to do with what the President was saying. But not long after, *New York Times* correspondent Frank Bruni watched her react the same way at a Budget Committee hearing. According to Bruni, Hillary "silently and reflexively shook her head and rolled her eyes almost every time one of the economists who were testifying mentioned Bush's tax cut."

The occasional slip aside, Hillary made a concerted effort to appear as if she was solidly behind the President in the immediate wake of 9/11. Two days after the attacks, she flew with the President to New York aboard *Air Force One*. (Chuck Schumer was reportedly peeved that Hillary took a seat next to the President and remained there the entire flight.) The state's two Democratic senators joined Mayor Giuliani and Governor Pataki at Ground Zero, where Bush talked to rescuers over a bullhorn. When someone in the audience yelled that he couldn't hear the President, Bush replied without hesitation, "I can hear you. The rest of the world hears you. And the people who knocked these buildings down will hear all of us soon!" The crowd then began chanting "USA! USA!"

Firefighters, police, and rescue workers swarmed over Bush, Giuliani, Pataki, and Schumer. But when Hillary stepped forward to share in the moment, they withdrew. "Nobody even wanted to shake her hand," said a firefighter on the scene. "She'd been bad-mouthing cops from the beginning, and the uniformed services didn't like that. So when she walked up and stretched out her hand, several rescue workers just folded their arms and turned away. Hillary got the message."

Hillary did not earn any extra points with police when, on October 14, her unmarked black Ford van blasted through a security checkpoint at Westchester Airport and headed for her waiting jet. Fearing that the van might contain terrorists, Westchester County police officer Ernest Dymond hurled himself against the car, pounding on the windows as he shouted for the driver to stop. Instead, the van containing Hillary, who was on her cell phone the entire time, barreled one hundred yards beyond the checkpoint before finally coming to a stop, injuring the officer.

Hillary never left the van. After Dymond was taken to the emergency room with an injured shoulder, his wife, Deborah, said she hoped Senator Clinton would call the hospital to check up on his condition. Either that or apologize. Hillary did neither.

It should have come as no surprise to Hillary, then, that many cops and firemen had come to despise her. Senator Clinton suffered her most humiliating rebuke on October 20, 2001, during Paul McCartney's nationally televised Madison Square Garden concert honoring the 343 firefighters and 87 police officers killed on 9/11. The five-hour concert, which also featured such performers as Mick Jagger, David Bowie, Elton John, Billy Joel, and James Taylor, raised more than $30 million for the victims' families.

The crowd cheered when Rudy Giuliani stepped onto the stage, and offered a warm welcome for the Democrats' leader in the Senate, Tom Daschle. But when Hillary walked to the microphone to introduce a short clip by comedian Jerry Seinfeld, the throng erupted in a chorus of boos. "Get off the stage!" yelled one firefighter in the front row. "We don't want you here!" Stunned, Hillary tried to be heard over the jeers by shouting into the mike. After a few minutes, she was forced to beat a hasty retreat. Although millions watching the live broadcast on VH-1 witnessed Hillary being booed off the stage, the tape was doctored so that during subsequent airings the jeers were disguised by general crowd noise.

The audience went wild, however, when firefighter Mike Moran, whose brother was among those killed, dared Osama bin Laden to "kiss my royal Irish ass." He felt similarly disposed toward Hillary. "When things are going well, people will sit there and listen to the kind of claptrap that comes out of her mouth," Moran said. But in rough times, he added, people were "not about to put up with her phony remarks. . . . She wants to get up and spew her nonsense—she doesn't believe a thing she says. She says whatever she thinks will fit the moment. I think it comes through, and in serious times, people don't want to stand for it."

Hillary later tried to shrug off the politically damaging incident. "I've gotten used to being in situations in political life," she said, "where that just happens sometimes." Calling the mortifying incident "part of the healing process" following the 9/11 attacks, Hillary said the police officers and firefighters "can blow off steam any way they want to. They've earned it."

In truth, Hillary was afraid that she might be booed again, and instructed her staff not to schedule any appearances in front of large groups of firefighters or police officers—at least not while tempers were still running high. She did deliver the eulogy at the televised funeral of New York City Fire Department Chaplain Father Mychal Judge, as well as appear at emotion-charged memorial services—also televised nationally—at Yankee Stadium and at Ground Zero. "These were no-brainers," said a member of Schumer's staff. "She *had* to go and be seen. It would have been political suicide not to go." But, with a final death toll of 2,749, there would be hundreds of funerals stretching out over the next several months—services that were not covered by the networks and were therefore of no particular political value to a senator with national aspirations.

Hillary attended not a single one. While Schumer showed up at a dozen funerals and Giuliani and Pataki paid their respects literally hundreds of times, Hillary avoided the possibility of being heckled by hunkering down in Washington.

Working with victims' families was another matter, or so Hillary repeatedly claimed, that consumed much of her time. When she heard that the author and journalist Steven Brill was writing a book on the aftermath of 9/11, she actually approached him during a ceremony at Ground Zero. Since Brill had been one of the Clintons' most ardent defenders during the Lewinsky affair and subsequent impeachment proceedings, Hillary indicated that she would cooperate.

Hillary began by saying that she had spent countless hours consoling the victims' families and doing anything she could to help. To back up that claim, her office provided what Brill called "an elaborate story, with an elaborate subtext of memos and phone calls—a long, long story." Brill was not surprised. "I think she has begun every statement she's ever made in her life about the families of the victims," Brill observed, "by saying she's spent innumerable hours with the families of the victims." Moreover, Hillary told Brill that it was she, and not Chuck Schumer, who was responsible for scoring billions of dollars in federal aid for New York.

Brill was surprised to discover that "none of it turned out to be true." Hillary's staff had given him "documents and phone calls and things like that which just plain never happened." A case in point was the family of 9/11 victim James Cartier. The Cartiers tried time and again to arrange a meeting with Hillary, only to be told that they needed to write a detailed memo first. "We don't meet with any families," Hillary's staff members said, "unless they write to us first and tell us what they want to meet about." The Cartiers finally gave up, convinced, Brill said, "that the only time families can meet with Hillary Clinton is if it's at a press conference."

By contrast, the Cartiers, who wanted to search for their relative's remains at Ground Zero and the Fresh Kills recovery site on Staten Island, had no difficulty arranging private meetings with both Schumer and Rudy Giuliani. The two men worked in concert to provide the Cartiers and all victims' families access to both sites.

Brill was astonished that, "in the cause of shooting down Chuck Schumer getting a bunch of pages in a book," Hillary would go to such trouble inflating her own record of service to victims' families. "What stunned me is that one person would try to steal away the credit from the other person, especially when everything I was hearing from the families is that Schumer" was more accessible. "It sort of takes your breath away."

"Brill's accusations are completely false and an obvious last-ditch effort to jump-start anemic book sales," responded Hillary's spokesman Philippe Reines, perhaps best known in New York political circles for reportedly streaking through the offices of his previous employer, unsuccessful New York mayoral candidate Peter Vallone, on a dare. "It's hard to imagine why Mr. Brill would choose to exploit such a horrible tragedy in this manner."

Chuck Schumer, for one, had no trouble at all believing it. Every time he stormed into her offices to accuse Hillary of upstaging him—on more than one occasion his shouting could be heard in the hallway outside Clinton's Senate offices—Hillary apologized and promised it would never happen again. But by early 2002, the escalating feud between New York's two senators was fast becoming, as Washington columnist Robert Novak put it, "the talk of Capitol Hill."

In the wake of 9/11, Hillary had other concerns. She was still haunted by her early advocacy of a Palestinian state—not to mention the fond embrace she gave Suha Arafat. In February 2002, Hillary grabbed headlines back in New York by traveling to Jerusalem and proclaiming her undying support for the state of Israel. Just as important, she took every opportunity to denounce Arafat. "Yasser Arafat bears the responsibility for the violence that has occurred," she declared during a visit to a pizza parlor where fifteen Israelis were killed by a Palestinian suicide bomber. "It rests squarely on his shoulders."

The fate of the Democratic Party in the 2002 elections often seemed to rest on Hillary's shoulders. The day she returned from Israel, Hillary helped raise $150,000 at a Manhattan fund-raiser for New York Congressman Greg Meeks. The day after that, she was back in Washington hosting a fund-raising dinner for Iowa's Democratic Governor Tom Vilsack. The following evening, she appeared at a $4.5 million fund-raiser honoring Vermont Senator Jim Jeffords, who bolted the GOP to become an independent and shifted control of the Senate back to the Democrats.

At the same time HILLPAC, which like other political action committees quieted down in the aftermath of 9/11, was back in full swing. The fund-raising operation was key to Hillary's building the kind of national power base she needed for a presidential run in 2008 or beyond. "Each and every one of the congressmen and senators and governors she helps out now will owe her big-time," said a onetime beneficiary of Hillary's largesse. "She's a very smart lady. She knows it'll pay off for her down the road."

While Hillary focused on making The Plan a reality, her husband traveled the world giving speeches—and contributing to an annual income that now easily topped $12 million. He had also, she would learn through the grapevine, reverted to his old ways.

Bill had once told one of the women in his life, "We're all addicted to something. Some people are addicted to drugs. Some to power. Some to food. Some to sex . . ."

Few women had more insights into Bill's character—and the reasons for his rampant womanizing—than his "Pretty Girl" Dolly Kyle Browning. Over the course of their off-again, on-again twenty-year affair, Dolly monitored Bill's wildly fluctuating weight and concluded there was an inverse correlation between his weight and his sexual activity.

According to Dolly, Bill was "an addictive personality" who, when he wasn't active sexually, replaced his admitted addiction to sex with his admitted addiction to food. His weight soared, Dolly said, whenever Hillary "had him on a tight leash or he was in the doghouse." Weight loss was "a sure sign that he's up to his old tricks." Over the three-year period after leaving the White House, six-foot-two-and-a-half-inch Bill would slim down from 225 pounds to a comparatively svelte 190.

For months, Bill's New York office had been posting notices in the political science departments at Columbia, New York, and Fordham universities advertising for students to intern at Clinton's Harlem offices. The criteria for judging applicants was exposed by three young women who mailed their résumés to Clinton's office and waited for a reply that never came. When the applicants called, they were told that the positions had been filled.

At that point, one of the women—a buxom redhead named "Katy"—dropped into Clinton's office in February 2002 and resubmitted her résumé, this time with a photo showing her in a tight sweater. According to her résumé, Katy had never gone to college, but she did say in her cover letter that she knew "enough to be deferential and follow directions. I keep my mouth shut when people who know what they're talking about are there. I listen. I'm smart. Please give me a shot."

Within three hours, Katy received a call from the former President's office asking her to come in as soon as possible. Her cover letter, said one of the staffers in charge of bringing candidates to Clinton, "really stood out."

Left to his own devices, Bill would, in fact, be linked in the press to more than a dozen women between 2001 and 2004. Reporters speculated about old friends like former Miss Arkansas Lencola Sullivan, whose Columbus Avenue apartment was only a quick jog from the former President's Harlem offices. Bill threw an impromptu party for Sullivan in his office when she married a Dutch security specialist. But even after she moved with her new husband to Amsterdam, the still-striking former beauty queen kept her Manhattan apartment and continued to see Bill whenever she was in town.

As for stunning Park Avenue socialites rumored to be involved with the ex-President, billionaire Revlon Chairman Ron Perelman's ex-wife Patricia Duff was not alone. Lisa Belzberg, thirty-eight-year-old wife of Seagram's heir Matthew Bronfman, became the object of speculation after she attended a Super Bowl party Bill threw in his Harlem offices. "There was a chemistry between them, no doubt about it," another guest later recalled. Bill, who reportedly pulled Belzberg toward him and murmured off-color jokes in her ear, apparently agreed. "She married a guy worth $6 billion," he boasted, "but she still likes to flirt with me."

By late February, Bronfman and Belzberg had separated. Over the next few months, Bill visited her home on several occasions, and was spotted cozying up to Belzberg at several charity events in Manhattan.

Belzberg would eventually reconcile with her husband, but Bill's plate was still full. Clinton was also allegedly meeting up with an unidentified blond woman on a weekly basis at New York's trendy Hudson Hotel, always arriving separately in the afternoon and spending a couple of hours in the hotel's penthouse suite. The pattern followed Bill's penchant for midday rendezvous with paramours in hotels and apartments. "It is absolutely his MO," said one.

At about the same time, he was spotted flirting with Ghislaine Maxwell, daughter of the late British press lord Robert Maxwell. She, in turn, brought Clinton together with mercurial supermodel Naomi Campbell in April 2002. Not long after, Campbell, who had just broken up with fifty-year-old Benetton tycoon Flavio Briatore, accepted Clinton's invitation to join him at the Austrian ski resort of Ischl. When she arrived from Milan, Bill, who was surrounded by other dignitaries attending the World Summit for Youth and Economic Development, did his best to appear surprised. "What's she doing here?" he asked in feigned disbelief before whisking her off to the town's exclusive Panorama Restaurant.

Campbell would also deny any romantic involvement with Clinton, but one eyewitness to the rendezvous marveled at how "totally comfortable they were in each other's company. Lots of laughing, heads together, whispering to each other. Obviously, a very warm relationship . . ."

Charlotte Dawson, the statuesque blond host of Australian television's top-rated *How's Life?* show, also admitted "dating" Bill when he visited New Zealand in June 2002. They were spotted rushing out of the Auckland Hilton together after Clinton delivered another $200,000 speech—this one launching the new BMW-7 series. After their secret late-night rendezvous, Dawson, the ex-wife of Olympic silver medalist swimmer Scott Miller, insisted Bill was "a really nice bloke, and very charming." But, she added in a rather unfortunate choice of words, "nothing went down." The man who introduced Bill to Charlotte, Australian promoter Max Markson, went on record as saying the former President spoke "very highly" of her following their evening together.

Another attractive young woman who did not deny that she "enjoyed dates" with Bill Clinton was twenty-nine-year-old, six-foot-tall actress, socialist, feminist, and former model Saffron Burrows. Burrows's marriage to fifty-three-year-old film director Mike Figgis (*Leaving Las Vegas*) was ending when she first began seeing the former President.

"I'm Bill's bodyguard," the British beauty liked to joke, admitting that he was a "close acquaintance" of hers. Burrows also conceded that, while Hillary was hard at work on Capitol Hill, she went out on numerous occasions with Bill in London and in New York. "He has," she said slyly, "a great sense of fun."

To make matters even more interesting, Burrows was an outspoken bisexual who went on record years before she met Bill saying that she had a crush on Hillary. "Bill found that rather funny" when she told him, Burrows said.

At about the same time, Hillary learned of Bill's questionable trip to Rio. It was enough, given the upcoming midyear elections in which she had invested so much time and energy, for Hillary to once again pull in the reins. "There was a showdown in Chappaqua," said one of their oldest friends in Arkansas. "Hillary basically told him that she didn't like what she was hearing and to cool it. She had done so much for him, she said, the least he could do was keep his pecker in his pants. Or words to that effect."

Despite constant speculation in the tabloid press that Hillary was on the verge of divorcing her husband, nothing was further from the truth. "What surprised me about Hillary Clinton to this day," said talk-show host and longtime Hillary supporter Rosie O'Donnell, was that "I assumed as soon as she was elected senator she would divorce. I am shocked that she did not. I told all my friends, 'As soon as she's in the Senate seat she will be a single senator.'"

But there was already heavy speculation that, since Hillary was far and away the most popular Democrat in the country, she would seek the presidency as early as 2004. Senator Clinton insisted that she had no interest in running for President—especially not against George W. Bush, whose poll ratings had soared in the wake of 9/11. But when she did seek the office, she would be running with a loyal spouse at her side. Hillary, who believed Al Gore had made a fatal error by distancing himself from the record of the first Clinton administration, intended to trumpet what she viewed as the Clintons' stellar domestic and foreign policy record. "Hillary sees herself as Eleanor to Bill's FDR," said a longtime FOH. "The Clinton legacy means a great deal to her. She is not going to do anything to compromise their place in history. No matter what she says about having to agonize over whether or not to stay with him during the Monica thing, she never seriously thought of divorce. And she never will. It's just not going to happen. Period."

Besides, New York's junior senator was too busy leading the charge against the Republicans. Hillary pounced on a report that the CIA had warned the Bush administration in August 2001 about Al Qaeda's plans to hijack American airliners. "I am simply here today on the floor of this hallowed chamber to seek answers to the questions being asked by my constituents," she intoned, pointing to the front page of the *New York Post*. "Questions raised by one of our newspapers in New York with the headline BUSH KNEW. The President knew what? My constituents would like to know the answer to that and many other questions. . . ."

Now clearly on a roll, Hillary also demanded to know "why we know today, May 16, about the warning he received. Why did we not know this on April 16 or March 16 or February 16 or January 16 or August 16 of last year?"

But she already knew the answer to that question. By this time, the White House had already explained in Senate briefings that the warnings did not include any mention of the possibility that the airliners themselves might be used as weapons against targets on the ground.

Still, the Senator made no effort to conceal the pleasure she took in hurling grenades at the party in power. "I never shy away," she made a point of saying, "from a fight."

Hillary received an unexpected gift in late June when, without explanation, U.S. Attorney James B. Comey closed the New Square clemency case. In the wake of 9/11, the investigation into Bill Clinton's decision to drastically reduce the prison terms of four Hasidic Jews who bilked the government out of tens of millions of dollars had simply lost steam—aided in no small part by President Bush's desire to "move on."

Similarly, the investigation into Clinton's decision to pardon fugitive financier Marc Rich would languish even after Rich's ex-wife Denise was granted immunity for cooperating with authorities. Oddly, the Bush administration would help the Clintons out again by refusing to release documents related to the pardons under the Freedom of Information Act. Invoking the doctrine of presidential privilege, the White House kept more than 4,340 pages of Pardongate documents under lock and key. U.S. District Court Judge Gladys Kessler, a Clinton appointee, would later uphold the Bush administration's argument that all records related to the Clinton pardons should remain secret.

George W. Bush had made it clear that he did not have the stomach for seeing *any* President dragged into court. In accordance with his boss's wishes, U.S. Attorney James Comey gave Bill and Hillary a pass.

One of Hillary's closest friends and allies would not be so lucky. Comey had been a protégé of Rudy Giuliani, who as U.S. attorney for the Southern District of New York made a name for himself prosecuting Mafia dons and the much-despised hotel queen Leona Helmsley. The same week he made the decision to drop the Justice Department investigation into New Square, Comey charged ImClone CEO Sam Waksal with insider trading—and implicated Hillary's longtime pal Martha Stewart in the process.

Friends of the domestic diva were convinced Martha Stewart had been sacrificed for Hillary Clinton. "Everybody was squeamish about going after a former First Lady," said someone close to the case. "But along came Martha. . . ."

For more than a decade, Stewart had been one of Hillary's most outspoken champions and contributed over $170,000 out of her own pocket to the Clintons and the Democratic Party. In 2000, she gave $1,000 to Hillary's Senate campaign, the maximum allowable to a single candidate under campaign finance laws. Like Denise Rich, Stewart hosted several high-profile events—including one at the Connecticut home of Miramax movie mogul Harvey Weinstein—that netted Hillary hundreds of thousands of dollars.

In return for her loyalty, Hillary invited Stewart to the White House on several occasions. Toward the end of their administration, the Clintons invited Stewart to bring television cameras into the Executive Mansion and film a segment for her series *Martha Stewart Living.* Stewart had even moved from Connecticut to Bedford in Westchester County, in part to be closer to her friends' Chappaqua home.

Their mutual friend Sam Waksal was also a major Hillary contributor to the tune of $63,000, all but $7,000 of that made to the Democratic Senatorial Campaign Committee. When he was charged by a federal grand jury with insider trading, obstruction of justice, and bank fraud, Hillary adamantly refused to join other Democrats who either returned Waksal's contributions or gave them to charity. But after heavy criticism, she reversed herself and donated the $7,000 she directly controlled to charity.

Hillary's kinship with Martha had everything to do with blond ambition. Both women believed they had suffered from the classic American double standard: a man who exuded confidence, strength, intelligence, and chutzpah was a leader. A woman who possessed these same qualities was more often than not branded a bitch. It was a designation that had, rightly or wrongly, been frequently applied to both Stewart and Senator Clinton over the years. "You know, with Martha and me," Hillary said of her friend, "it's kind of a mutual admiration society."

However strong the bond between them, Hillary's relationship with Martha would be exceedingly problematic. The day after Sam Waksal's arrest, Hillary canceled a $1,000-a-head Democratic fund-raiser Martha was throwing at the Manhattan headquarters of her Martha Stewart Living Omnimedia empire. Hillary's oft-used excuse for the sudden and inexplicable change in plans: "Scheduling problems."

Hillary had decided to hold on to the $1,000 Martha gave her for the time being—"she has not been found guilty of any wrongdoing," Senator Clinton argued—and went so far as to call her friend with words of encouragement. "She was one of the first people to call me," Stewart later recalled, "and very nicely say, 'You know, you just have to hang in there. It's the process.' "

Once word of the phone call was out, however, Hillary began circling the wagons. The senator simply "made a call to a friend," Hillary's spokeswoman Karen Dunn said, "and has not commented on the ongoing investigation." Was Hillary supportive? "I wouldn't say either way," Dunn replied.

Martha Stewart's once-tidy world would come undone over the next two years. Briefly a billionaire after her corporation went public in 1999, she lost control of her company and hundreds of millions of dollars as the federal case against her dragged on. Many believed Stewart was being unfairly prosecuted because, like Hillary, she was a powerful, ambitious, often abrasive woman in the male-dominated world of business. The very same qualities that made Hillary an object of fear and loathing now threatened to put Martha Stewart behind bars.

After that early phone call, Hillary was nowhere to be seen. At first, according to friends of both women, Martha understood that Hillary could ill afford to be tied to a Wall Street scandal. There was little tolerance for corporate greed following the horrific collapse of Enron and WorldCom. So, as she plotted her own legal strategy, Martha resigned herself to the fact that she wouldn't be hearing from either Clinton. "I think she felt it wasn't important that Hillary gave her the silent treatment in the beginning," said one doyenne of Manhattan society. "Martha is very savvy politically, and she didn't want to harm Hillary's career in any way. 'I think Hillary would make as good a President as Bill—better, actually,' she once told me. But I know she was confident Hillary would back her up if things got really tough."

Meantime Hillary, who still ordered Secret Service agents to carry her bags as she shuttled between Washington and New York, continued to raise millions for her fellow Democrats running for office in 2002. Since taking up residence there eighteen months earlier, Senator Clinton had held no fewer than twenty-eight fund-raisers at Whitehaven, her mansion off Embassy Row. On two separate occasions, she held back-to-back receptions on the same night. Whitehaven had become, the *New York Times* declared, "a conveyor belt of fund-raising dinners and receptions that Democratic candidates are clamoring to climb aboard." Hillary was doling out more money to Democrats across the country than any other senator, methodically shoring up support for her own run in the more distant future.

Giving the keynote address that July to the centrist Democratic Leadership Council, Senator Clinton eclipsed all of the 2004 presidential hopefuls in attendance—including Massachusetts Senator John Kerry, North Carolina Senator John Edwards, Connecticut Senator (and former vice presidential candidate) Joe Lieberman, and House Democratic Leader Dick Gephardt. In recent months, Hillary had held off on attacking the current occupant of the

White House. Under Bush's leadership, Operation Enduring Freedom had brought an end to Taliban rule in Afghanistan, and the U.S. was aggressively prosecuting its war on terrorism at home and abroad. But the declining economy and a rash of corporate scandals—Enron chief among them—had left W vulnerable, and Hillary seized the moment.

Without mentioning Bush by name, Hillary made the all-too-familiar populist argument that Republicans were friends of the rich and special interests. She also implied that the Bush administration, which was aggressively prosecuting several insider-trading and accounting-fraud cases, was somehow responsible for the rise in corporate corruption. The Republicans' attitude toward insider trading was, said Hillary, "Don't ask, don't tell . . . GOP used to stand for Grand Old Party, but more and more, it stands for Gloss Over Problems and pretend nobody notices!" The crowd leaped to its feet, applauding and cheering wildly.

"She's more effective than pretty much anybody but her husband," gushed council member Jack Weiss. "She's just got it." Maine State Representative Lisa Tessier Marrache echoed the sentiments of many party rank-and-file members. "I wish she would run," Marrache said. "She's very charismatic."

Hillary's give-'em-hell speech was especially audacious, since she had gladly accepted large contributions from such corporate train wrecks as ImClone's Sam Waksal, Martha Stewart, WorldCom, Enron, and the accounting firm of Arthur Andersen. "That," she explained, "was in the past. . . ."

Hillary now actively encouraged her fellow Democrats to fire away at Bush's integrity; in closed-door meetings she urged party operatives to dredge up Bush's sale of Harken Energy stock in 1990. Back then, the Securities and Exchange Commission had looked into the transaction—which was executed shortly before Harken's stock took a nosedive—and Bush's failure to file the proper paperwork. The SEC ruled at the time that the case merited no further action. Hillary, apparently giving little thought to her own precarious position and that of her friend Martha, felt Bush was particularly "vulnerable" in the current anti-big-business climate.

That summer, Hillary also had to deal with the news that her husband was in the midst of negotiations with NBC and CBS for his own syndicated daytime talk show. Their old Hollywood pal Harry Thomason was handling talks with network executives, demanding that the former President be paid $100 million to host the show for two years. After NBC backed out, Bill reportedly dropped his price to around $30 million a year.

There was the inevitable spate of headlines when news of Clinton's talk-show negotiations were leaked to the press: VIEWERS WOULD SNUBBA BUBBA, BOOB-TUBE BUBBA, and BUST-SEE TV: BILL SHOW A DUMB IDEA. "You were leader of the Free World!" Rosie O'Donnell protested. "Don't do a talk show, you moron! If he really does a talk show, I'm becoming a Republican. I swear to you—fully Republican. You're going to see me all over the country campaigning for the Bushes. If it is true, I want to see if I can get any money I donated to him back!"

Bill's wife was mortified. "Hillary was just really pissed off," said a friend from Little Rock who had served in the Clinton administration. "Here she is running around the country trying to mobilize the party, and Bill is acting like he wants to be the next Jerry Springer. Hillary thought it was beneath him and it made her look bad, so she told him to knock it off." Eventually, he would, but only because no network was willing to match his nine-figure asking price.

On Capitol Hill, Hillary continued to placate the Senate's entrenched male leadership. At their regular Tuesday caucus meetings, said Louisiana Democrat John B. Breaux, "the thing that has most impressed me is sitting down and watching her get up to get coffee and ask the other senators whether she could bring them back some coffee. She aggressively avoids the spotlight and intentionally holds back, while most of us try to do the opposite."

On the stump, however, Hillary was once again the frequently ferocious partisan. "The stakes are so high in this election," she said in late October, "I just have to impress upon you how critical this is. If we were to lose the Senate, there would be nothing standing between the Republican leadership and confirming the most extremist judges, in rolling back environmental regulations. And certainly, you might as well say goodbye to fiscal responsibility."

At private fund-raisers, Hillary's attacks on Bush and the Republicans were described by one *New York Times* reporter as far more "brutal and alarmist" than what she was willing to say in public. "We're the party of the people, and they're the party of the rich and the special interests—it's really that simple," she declared at one small gathering. At another, she said W was "much worse than his father, because he owes his soul to the far right." She was also convinced that Vice President Dick Cheney, for whom she had long had a visceral dislike, was "really running the show."

As the chances of a Democratic sweep began to fade, it became clear that Hillary—still one of the most polarizing figures in American politics—may have done as much for the Republicans as she had for her own party. In numerous contests across the country, the GOP ran television campaigns trying to link Hillary to various Democratic candidates.

In many cases, the ploy worked. So, too, did George W. Bush's own campaign blitz across the country on behalf of Republican candidates. When the dust had settled, Democrats had suffered a stunning off-year defeat as Republicans regained control of the Senate and extended their lead in the House. Fortunately for Senator Clinton, the historic GOP victory was chalked up to George W. Bush's continuing popularity, and not to an anti-Hillary backlash.

Undaunted, Hillary now focused her attention on 2004, and how that election might affect her own planned run for office in 2008. On *Meet the Press,* she repeated her assertions that she was "110 percent certain" she would not run for President during the next election. But when asked about 2008, she followed her husband's example and parsed her words carefully. While not ruling it out explicitly, she said she had "no plans" to run in 2008.

"Hillary will be sixty-one in 2008, and even if she waits until 2012, she'll only be sixty-five," said one Democratic strategist. "Reagan was seventy when he was elected, and half the country will be baby boomers around her age anyway. So she has time." By the end of the year, it appeared that Hillary's denials had fallen on deaf ears. A CNN/*Time* magazine poll indicated that 30 percent of registered Democrats would vote for her for President, compared to just 13 percent each for John Kerry and Joe Lieberman.

One Democrat who had removed himself from the running, Al Gore, had done so in part because Hillary had made it clear she would not endorse him—at least not before the primaries. "If I had a really, really good friend—as you've described Al Gore to me," Chris Matthews asked Hillary on WNBC's *Hardball,* "a really, really good friend, and he was telling me I think I'm going to run for president . . ."

"He hasn't said that to me," Hillary replied.

When Matthews suggested that a real friend wouldn't have to be asked, that she would say "I'm with you, buddy, all the way" beforehand, Hillary shrugged.

"He hasn't talked to me about it. . . . No, you know, Chris, I don't endorse in Democratic primaries."

"Al Gore got the message loud and clear," said a former member of the ex–vice president's staff. "You know, it's really a case of 'with friends like that.' Hillary is the most influential person in the party right now, and she was letting everyone know she had no faith in him."

When John Kerry stepped down as chair of the key Senate Democratic Steering and Coordination Committee to run for President, it was Hillary who was tapped to take his place. The little-known committee was charged with putting out the party message to officials all over the country as well as approving appointments to Senate committees. Then she was given a seat on the powerful Senate Armed Services Committee. This gave Hillary a pulpit from which to expound on Bush's foreign policy as he prepared to wage war on Iraq.

Senator Clinton shrugged when it was pointed out that both jobs perfectly positioned her to expand her power base and influence for a run in 2008—if not sooner. "I just want," she said, "to be effective."

Hillary went back on the attack in January 2003, now charging that Bush's Homeland Security plan was a "myth" and that Americans were more vulnerable to terrorism than ever. "Our vigilance has faded at the top, in the corridors of power in Washington," she said, "where leaders are supposed to lead." Once again, the senator said Bush put the interests of the wealthy above the security of ordinary citizens. "Will ending the dividend tax keep a dirty bomb out of New York harbor?" Hillary also accused the President of turning a blind eye to the nuclear threats in Iran and North Korea, and of being "fiscally irresponsible" by cutting taxes.

Nonetheless, Hillary voted with the majority of Democrats and Republicans to grant the President congressional authority to wage war on Iraqi dictator Saddam Hussein. "It was the hardest decision," she said, "I've ever had to make." The U.S.-led offensive to oust Hussein got under way on March 19, 2003, and for the next several weeks, while American troops pressed toward Baghdad, Hillary remained uncharacteristically silent.

By April, however, Hillary and Bill seemed to be everywhere—speaking at seminars and luncheons, doing television and radio interviews, schmoozing with the party faithful at countless events from coast to coast, and making the kinds of remarks that were certain to keep them center stage. Asked if his wife was going to run for President, Bill started tongues wagging by suggesting that she'd "make a better vice-presidential candidate" in 2004.

This was no accident. The Clintons' former adviser Dick Morris claimed there was a "conscious effort going on by the Clintons to distract attention from the current field of candidates. They do not want a Democrat to win in '04." Indeed, as part of their effort to "trivialize" the other candidates, the Clintons had refrained from giving money to any of them.

Susan Estrich, a longtime Democratic Party strategist and close ally of the Clintons, conceded Bill and Hillary "suck up every bit of the available air. Nothing is left for anyone else. They are big, too big. That's the problem. . . . The 2004 candidates need a chance to get some attention. Could somebody please tell the Clintons to shut up?"

Estrich and the rest of the party would be swept away in a tide of Hillarymania with the June 2003 release of her $8 million memoir, *Living History.* Completed on schedule (this time with the help of six ghostwriters), the book landed Hillary on the cover of *Time,* on front pages everywhere, alongside Barbara Walters on ABC, and at number one on the *New York Times* bestseller list.

Much of *Living History* was devoted to Hillary's midwestern upbringing, her involvement in Bill's campaigns, her social causes (health care, children's rights, welfare reform), and her travels abroad. Yet what really captured the public's attention was Hillary's mind-boggling claim that she, too, was shocked when her husband confessed about his affair with Monica Lewinsky.

For the most part, Hillary's autobiography had less to do with living history than with rewriting it. Still lashing out at the "vast right-wing conspiracy" supposedly aimed at bringing down the Clintons, Hillary defended her husband even as she vividly described his "stinging betrayal" of their marriage.

Critics, for the most part, were incredulous. "*Living History* is neither living nor history," wrote the *New York Times*'s Maureen Dowd. "But like Hillary Rodham Clinton, the book is relentless, a phenomenon that's impossible to ignore and impossible to explain."

Whatever its literary merits, *Living History* proved to be a potent political tool. By stressing her domestic and foreign policy credentials, and portraying herself once again as the wounded but loyal wife, the only First Lady to win elective office was offering up what amounted to a presidential résumé. "What she has done," said New York–based Democratic consultant Hank Sheinkopf, "is create a national constituency for a newly defined Hillary Clinton." Concurred journalist Joe Klein: "This is the memoir of an active—and very ambitious—politician. The Senator is looking to augment her political viability."

The result was that Hillary, still proclaiming her desire to see some other Democrat elected to the White House in 2004, remained with her feet planted squarely center stage. "She has commanded more attention," said another Democratic strategist, Philip Friedman, "than the nine Democratic presidential candidates combined, she has given her version of a scandal that involved her family, and she has begun to move on to a posture as a national leader in the party." Another Democratic consultant added that the candidates "must be going out of their minds today! They can't even get on Page A27, but Hillary's on the front page of newspapers all over the country."

Then there was the money. Tucker Carlson, the conservative half of CNN's *Crossfire,* had vowed to eat his shoes if Hillary's book sold a million copies. When it did after just one month, Hillary showed up on the *Crossfire* set with a chocolate cake in the shape of a shoe. "It's a right-wing wingtip," she announced.

While promoting the British edition of the book (*Living History* reached number one in England, France, and Germany), Hillary uttered the same line over and over to describe the Lewinsky affair: "Well, it should have remained private and personal but it was forced into the public for partisan political purposes which I found deplorable . . . these people were willing to destroy anyone in order to end my husband's presidency." She also repeatedly denied that she had any plans to run for President. "What about 2008?" one interviewer asked. Hillary smiled and answered that "2008 is an eternity in American politics." However, she added, "You never know what might happen." She also volunteered that if she were to be elected President, she wanted her husband to be called "First Mate."

By late August 2003, there were strong indications that the book had gone a long way toward rehabilitating Hillary's image. Before its publication, someone had pointed out to Hillary that she remained unpopular in large parts of the country. At the time, Hillary grinned broadly and replied, "That's because they don't know me!" Now, in the wake of her nationwide blitz to promote *Living History,* a Gallup poll showed that Hillary's favorable rating had gone from 43 percent to 53 percent.

To maintain momentum—and keep her fellow Democrats in a perpetual state of bewilderment regarding her intentions—Hillary instructed aides to keep posting e-mails on her official Web site from fans exhorting her to toss her hat into the ring. Asked why she was allowing hundreds of such "Hillary for President" e-mails to be posted, Hillary answered matter-of-factly, "Freedom of speech, I guess."

No politician made more efficient use of the Internet, in fact. "Have you picked up the paper lately or clicked on a cable channel to find someone saying the most outrageous things about Hillary Rodham Clinton?" asked the FriendsofHillary.com home page. "We have! It really steams us that some people would twist the truth or use such hateful language to attack a Senator who is working hard to make life better for the people of New York and the nation. The right wing is even angrier because they've been unable to stop Hillary with their vicious personal attacks. We're fighting back! Become a HILLRAISER today!"

Another FOH-sponsored site belongs to "Hill's Angels," which zeros in on fund-raising. "While Hillary is fighting for the values and policies we care about," reads the introduction to the Web site, "the right wing is waging a personal attack against her." Hillary personally signed off on the copy, hence the many references to her "right wing" critics. Meanwhile, Hillary received an endorsement that, given the rocky state of Franco-American relations during the Iraqi War, she might just have well done without. Bernadette Chirac, wife of French President Jacques Chirac, said on French television that "a lot of women hope that one day she will run for the presidency of the United States and that she'll win."

Buoyed by her rising poll numbers—numbers that still showed her twenty points ahead of any declared candidate—Hillary went back on the offensive. This time, she charged that the White House had pressured the Environmental Protection Agency to downplay the fact that toxins were swirling in the air after the collapse of the Twin Towers on 9/11. "I don't think any of us ever expect to find out," she said, "that our government would knowingly deceive us about something as sacred as the air we breathe. . . ." She then called for Senate hearings to investigate what she was now calling a "cover-up." Hillary said she could "see no other way to get the administration's attention."

Later, the senator would up the ante by threatening to block the confirmation of outgoing Utah Governor Mike Leavitt as the new head of the EPA. In a statement that revealed more about the way the Clintons operated than Hillary may have intended, she said, "I know a little bit about how White Houses work. I know somebody picked up a phone, somebody got on a computer, somebody sent an e-mail, somebody called for a meeting, somebody in that

White House probably under instructions from somebody further up the chain told the EPA, 'Don't tell the people of New York the truth.' And I want to know who that is."

Hillary would keep hammering away at Bush's record, confident that not one of the Democrats lining up for the nomination had any chance of beating him. Six months after American troops entered Iraq, support for the war was hovering around 63 percent. Polls also showed that, for the time being, Bush would easily recapture the White House, leaving the field wide open to Hillary in 2008.

The Clintons continued to ignore Susan Estrich and other party operatives who begged them not to hog the limelight. When CNN's Judy Woodruff asked if Hillary wasn't guilty of distracting attention from her fellow Democrats, Hillary shrugged. "Well," she said, "I don't see that at all." Then she argued that the party's best hope was to keep reminding Americans of "the Clinton Administration and the difference it made in the lives of so many Americans."

Her hollow denials notwithstanding, Hillary kept dangling the possibility in front of the party faithful that she would jump into the race at any minute. Bill was her co-conspirator in this ongoing effort to keep the rest of the field off balance. On a Sunday in early September, the Clintons hosted a dinner for 150 major donors—people who had contributed $100,000 or more to the Clintons over the past twelve months—at their home in Chappaqua. At one point over cocktails, Bill said the Democrats had only "two stars"—his wife and General Wesley Clark, the retired NATO commander both he and Hillary had secretly approached to enter the fray. Later, Hillary told her dinner partner, Gristedes supermarket chain CEO John Catsimatidis, that "we might have another candidate or two jumping into the race." Said Catsimatidis, "I didn't get the impression she had pulled the trigger in her mind about whether or not to run." He was "left with the impression that there's always a possibility."

To another guest, Hillary pointed out that she would be needing additional money—lots of it—for a campaign "somewhere down the road." One major backer of the Clintons later assured her fellow guests that they "were not hallucinating." Hillary's playful attitude, as well as her husband's provocative asides, had left everyone scratching their heads.

Ten days later, Wesley Clark entered the race with a campaign team that included many familiar names from the Clinton administration. Clark announced at the time that Senator Clinton had promised to sign on as co-chair of his campaign—a statement that Hillary's camp refused to confirm or deny at first.

Like Bill, Clark was an Arkansas-bred Rhodes scholar who zoomed to the top of his chosen profession at an early age. Clark had led the successful military operation in Kosovo, but nevertheless was reportedly relieved of his command by envious superiors at the Pentagon. Then-President Clinton went along with Clark's sacking at the time, though he later claimed to have known nothing about it. Now, four years later, it looked as if Bill and Hillary, who had been chatting up General Clark as a possible candidate for months, were about to make amends by getting behind a candidate they thought could win.

Or were they? On the eve of Clark's announcement, Bill told an audience in California that he thought the general would do fine in the short run. "Whether he can get elected president," Clinton added, "I haven't a clue." Instead, he again touted Hillary's chances of winning back the White House for the Democrats. "I was impressed at the state fair in New York, which is in Republican country in upstate New York," he said, "at how many New Yorkers came up and said they would release her from her commitment [to serve out her Senate term] if she wanted to do it."

This was precisely the same grassroots argument that Clinton himself had used in 1992 to justify breaking his promise to Arkansans that he would serve out the remainder of his final term as governor rather than run for President. Hillary had urged her husband to break his pledge not to run back then, and she had no trouble reneging on her own Shermanesque declaration in 1997 that she would never seek elective office.

At this point, looking at a watered-down field of ten candidates—none of whom came within fifteen points of George W. Bush in the polls—Bill was pushing Hillary harder than ever to get into the race. He argued that Bush was now especially vulnerable: Conditions in Iraq, where American occupation troops were under fire from insurgents, were chaotic. None of the "weapons of mass destruction" that had been used to justify the war had yet been found. And, most important, unemployment was up and the economy was still struggling.

"He is really after her to do it," one of the Clintons' closest Arkansas confidantes said at the time. "But the bottom line is that she isn't sure Bush can be beat. She trusts her own instincts more than she trusts Bill's."

As it turned out, Bill was not the only member of their tight-knit family urging Hillary to run. Chelsea had returned from her studies at Oxford—where anti-Bush feelings ran high—harboring a bitterness toward the President that transcended her parents' own disdain. Chelsea, now ensconced in New York and earning six figures as a consultant with McKinsey & Company, tried to convince her mother that she was the only person who could "rescue" the country from the Republicans.

By late September, pundits were speculating that the Clintons had backed Wesley Clark simply to foil the then front-runner, Vermont Governor Howard Dean, and set the stage for a last-minute entry by Hillary. She dismissed this as "an absurd feat of imagination, I guess. . . ." In fact, she was quick to point out, Hillary and Bill had not endorsed Clark. To drive home that point, the former President phoned three of the candidates and assured them neither he nor his wife was endorsing anyone.

On September 24, it was clear that Hillary had finally made up her mind to stick with her original plan and focus on running for President in 2008. No longer indulging in playful asides, she told some sixty journalists at a Washington breakfast meeting that she was not running in 2004—not once but a dozen times.

She was not relinquishing her role, however, as Democratic Party hit man. For an hour, Hillary blasted Bush's "shocking failure of leadership . . . I am convinced totally that four more years of this administration would be an overwhelming setback for our country. And," she was careful to add, "I will do everything I can to elect whoever emerges from this process."

For Howard Dean, John Kerry, North Carolina Senator John Edwards, Joe Lieberman, and the rest, it might have been enough that Hillary now seemed to be bowing out once and for all. But that did not mean she would cease being a factor in the 2004 elections. Of all the Democratic candidates, it was Kerry who feared Hillary the most. The Massachusetts senator had been allied with the Clintons both personally and professionally for over two decades, and was well aware that Hillary had plans to recapture the White House. "John doubted that she'd jump in," says a Boston politico who has known Kerry for over a quarter century. "But he sure as hell knew she didn't want another Democrat to beat Bush, no matter what she said."

Kerry blanched during CNN's "America Rocks the Vote" special in early November 2003, when moderator Anderson Cooper read an e-mail from a viewer asking Kerry what he thought about polls showing Hillary Clinton 20 percent ahead of "all you guys."

"Well, it all depends on which poll you look at," Kerry replied. "I saw a poll the other day that showed me about 15 points ahead of her."

Pressed to identify the mystery poll that showed him ahead of Hillary, Kerry came up dry. After a few days, an aide to the Massachusetts senator conceded that Kerry "misspoke." To be sure, polls now showed him trailing Hillary by a whopping 35 percent. Hillary shook her head when a staffer told her about Kerry's quasi-admission that he had lied. "Poor John," she sighed.

As much as Hillary's staff enjoyed watching Kerry and the rest of the Democratic field squirm, they took even more pleasure in the predicament faced by Hillary's longtime nemesis Rush Limbaugh. That autumn the conservative commentator, who had advocated stiff prison sentences for drug offenders, confessed that he was battling an addiction to the painkiller OxyContin. Limbaugh's admission would trigger an investigation into how the drug was obtained.

"You have got to be kidding!" Hillary squealed when she first learned of Limbaugh's drug use. She was careful not to gloat in public, but according to one longtime resident of Hillaryland, "you could tell she enjoyed watching Rush twist slowly in the wind."

Hillary had officially removed herself from consideration in 2004, but that did not mean she intended to step aside. A major event of the primary season was the Iowa Democratic Party's Jefferson-Jackson fund-raising dinner, where candidates made their case to 7,500 party activists. What was said at the Jefferson-Jackson Dinner could well determine the outcome of the Iowa caucuses the following January.

By agreeing to moderate the dinner, Hillary made it clear she was not about to relinquish her position as the Democrats' superstar. While the six candidates who chose to attend smiled through gritted teeth, Hillary, looking demure in lace collar, lace cuffs, and Barbara Bush pearls—an outfit she would recycle at several high-profile events in the coming months—railed against Bush and the GOP. Outside the hall, demonstrators waved placards urging Iowans to vote for Hillary in the Iowa caucuses. "Senator Clinton's ability to outshine and overshadow the Democrats running for president," observed Republican National Committee spokesman

David James, "is further evidence of the weakness of the Democratic field."

With her focus on running in 2008, Hillary could campaign for the party's 2004 nominee secure in the knowledge that he had little chance of winning the national election. Two weeks later, she again felt free to say to a foreign audience what she would never say at home. When a reporter for the German magazine *Bunte* said people were disappointed that she wasn't running for President, she replied, "I know. Well, perhaps I'll do it next time around."

One area where Hillary sought to distance herself even further from the pack was foreign policy. Unlike the then front-runner, Howard Dean, who had always opposed the U.S. invasion of Iraq, Hillary had taken a decidedly hawkish stance at the outset. That did not keep her, however, from criticizing the Bush administration for its management of the war. She claimed to be "bewildered, surprised, disappointed by the failure of the administration to create conditions for greater international support," and slammed Bush as "the first president who has ever taken us to war and cut taxes at the same time. I view that as incredibly irresponsible. But the happy talk continues. . . ."

John Kerry had also voted to give the President authority to send troops to Iraq, and was now taking every opportunity to bash Bush for the government's failure to find the elusive WMDs (weapons of mass destruction) that had been used to justify the war against Saddam Hussein in the first place. Of all the candidates, in fact, only Kerry, a senior member of the Senate Foreign Relations Committee and the Senate Intelligence Committee, boasted a foreign policy background that could rival that of Bill Clinton's well-traveled co-President.

So when Hillary announced she would be paying a Thanksgiving holiday visit to U.S. troops in Afghanistan and Iraq, Kerry was understandably—according to a former Democratic congressman from his state—"fucking mad. It was All Hillary All the Time. Everything she did was designed to keep things static, so the guy with really no chance of beating Bush—Howard Dean—would get the nomination."

"I am honored to have the opportunity to spend the Thanksgiving holiday visiting with the men and women who have put their lives on the line for all Americans," Hillary said in a press release that sounded . . . presidential. Before she left, Hillary went a step further and aired a televised message to the nation paying tribute to the troops in Iraq and Afghanistan for their sacrifice. "I want to thank all of our brave men and women who are serving our country," intoned Hillary, seated in front of an American flag and sounding more like a commander in chief than just one of a hundred senators. "And I want to thank and honor the service of all of you who have been wounded during all of these conflicts in Afghanistan, Iraq and elsewhere defending our country, fulfilling your mission and making us all very proud."

Accompanied by Rhode Island Democrat Jack Reed, a fellow member of the Senate Armed Services Committee who had voted against sending American troops to Iraq, Hillary first met with Afghan President Hamid Karzai in Kabul's stately presidential palace. Later, they cut to the head of the line in the linoleum-floored army mess hall at Bagram Air Base and loaded up their cardboard trays with turkey, stuffing, mashed potatoes, cranberry sauce, yams, and pumpkin pie.

After she left the dining hall, Hillary learned the stunning news: President Bush had made an unannounced, lightning-swift trip to Iraq, where he served Thanksgiving dinner to American troops before chowing down with them. For security reasons, the top-secret mission remained under wraps until the President was safely on his way home. Back in the U.S., video of the President's surprise visit—and his emotional reaction when the troops leaped to their feet to cheer him—preempted virtually all television programming for hours. It was a major news event, and an undeniable coup for the President. Hillary put on a dignified front in front of the troops, but when she thought she was out of earshot she mumbled her reaction to Bush's preemptive visit: "Son of a bitch."

The White House dismissed the notion that Bush had undertaken the trip to undercut Hillary; the President's Thanksgiving surprise had actually been in the works for months. "But," said a lifelong friend from Texas, "skunking Hillary made it that much sweeter." When he asked the President about Hillary, Bush feigned surprise. "Was she there, too?" he asked.

Trumped by W, Hillary nonetheless soldiered on. Before embarking on her whirlwind tour of the region, she had said that its principal purpose was to boost morale. But after arriving in Baghdad on the President's heels, her comments were something less than encouraging. She told the troops that while "Americans are proud" of them, "many question the administration's policies . . . the obstacles and problems are much greater than the administration usually admits to."

Echoing a phrase that hearkened back to the Vietnam era, Hillary also stressed that "it's no longer enough for our military, the most powerful in the history of the world, to win the battle—they have to also win the war for hearts and minds."

Hillary, who accused Bush of being "obsessed with Saddam Hussein for more than a decade," also called for more UN involvement—and for more American soldiers to be sent to Iraq. "The Pentagon tried to make do with as few troops as possible," she said, "as light a footprint as they could get away with. Now we're playing catch-up."

Undeterred by the fact that Bush's Thanksgiving surprise had shoved her off front pages across the country, Hillary went to all three network Sunday-morning news shows to report on her trip to Afghanistan and Iraq. When she was asked why she would tell troops stationed in Iraq that Americans were questioning the war and that its outcome was "not assured," Hillary froze. Once again, she reverted to the "vast right-wing conspiracy" argument to defuse legitimate inquiries from the hosts of NBC's *Meet the Press,* CBS's *Face the Nation,* and ABC's *This Week with George Stephanopoulos.*

"You know," she told Tim Russert of *Meet the Press,* "I find this so interesting that this has now become an issue, and largely fueled by a lot of the talk shows and the other sort of right-wing apparatus." On *Face the Nation,* she went so far as to deny that she had ever made the statements questioning U.S. policy while visiting the troops. She now dismissed the controversy over her remarks as "the latest flaming charge from the right wing."

More than ever, Hillary felt under siege. She was sitting in the kitchen of the Chappaqua house listening to a radio talk show when someone called in to say the Clinton administration had barred members of the military from entering the White House wearing their uniforms. She picked up the phone and called the station to deny the story.

"Literally, I have been accused of everything from murder on down," Hillary told Brian Lehrer's listeners on WNYC Radio. "And it's hurtful and personally distressing when it first happens but when it continues . . ."

The next surprise came not from the right but from the left, and it would take place right under Hillary's nose. Just down the street from Bill's Harlem office, Al Gore announced that he was endorsing Howard Dean. The former Vice President praised Dean for opposing the war in Iraq ("a catastrophic mistake," Gore said) and declared that the centrist Democratic Party that had been dominated by the Clintons for over a decade needed to be "remade" as "a force for justice and progress and good in America."

Hillary viewed Gore's endorsement of Dean, delivered deep in the heart of Clinton country, as nothing less than a betrayal. When asked if she agreed with Gore's claim that the party needed to be pulled back to the left, Hillary offered a terse reply: "No."

To be sure, the endorsement cost Gore nothing. Like the Clintons, he was convinced that Dean was too strident and too unknown to beat Bush. But Gore needed to seize control of the party from the Clintons for his own run in 2008, and to accomplish that he wanted Dean's energized young activists in his corner.

"This was not Al Gore taking a shot across Hillary Clinton's bow," said University of Virginia political science professor Larry Sabato. "This was him putting one right into the solar plexus of both Clintons."

Hillary got the message. But it did not radically alter her view of Clinton's once-loyal VP. By this time, said one New York State party official, Hillary "already thought Al Gore was an ungrateful putz."

There were times she felt the same way about her husband. With Bill almost constantly on the road making speeches, the Clintons seldom saw each other more than once or twice a month. Hillary volunteered that they spoke on the phone daily, and while they needed to stay in contact to strategize and maintain their hold on the Democratic Party apparatus, they often went a week or longer without speaking to each other.

Yet in January 2004, the phone lines between Chappaqua and the Embassy Row house were buzzing over gossip that Bill was now involved with another attractive young blonde—Canadian billionaire Belinda Stronach. At thirty-seven, Stronach was president and CEO of Magna International, the $13 billion Ontario-based auto-parts empire founded by her father, Frank Stronach.

Bill had met Belinda three years earlier, over a round of golf at the Magna course outside Toronto. Since then, they'd gotten together on numerous occasions. In September 2002 Bill and Belinda, invariably swathed in designs by Prada, Armani, and Gucci, attended a birthday party in Toronto for rocker Ronnie Hawkins. They then ducked out early to share a quiet dinner at one of Toronto's tonier restaurants, Truffles. The following year, they dined together again at the Democratic Governors' Conference in Baltimore and at a California fund-raising event before meeting up at the Preakness Stakes—Belinda's father owned the legendary Maryland racetrack—in May 2003.

That same month, Belinda divorced her second husband, Norwegian Olympic speed skater Johann Olav Koss, sparking what the *National Post* of Canada called a "wildfire of rumors in Toronto society" concerning the cause of the breakup. The paper went on to say that the rumors "all start and end with one word: Bubba."

In November 2003, Stronach was being given a humanitarian award at Toronto's Beth Shalom Synagogue when suddenly Bill's voice came over the intercom. Hillary's husband was just calling—from China—to congratulate Belinda, and to tell her how sorry he was that he couldn't be with her.

By this time, Bill had taken an active interest in Belinda's political career. At Bill's urging and with his help, she was now planning to take over as leader of Canada's Conservative Party—a move that would make her prime minister if her party won the next general election.

Bill and Belinda were spotted skiing together in Aspen in January 2004, and reportedly made plans to see each other again two weeks later at the World Economic Forum in Davos, Switzerland. It was at this point, say friends of both Clintons, that Hillary called Bill and voiced her concerns.

Belinda, meanwhile, insisted through her corporate spokesman that there had never been anything the least bit romantic between her and the former President. But friends of the glamorous young woman now being hailed as "The 'It Girl' of Canadian Politics" allowed that Belinda was "intrigued" by "his charisma and brainpower." Nevertheless, she abruptly canceled her trip to Switzerland, leaving Bill to attend the conference alone.

There was ample cause for alarm on Hillary's part. At this point, while they struggled to determine the best way to move forward with The Plan, the Clintons did not need more headline-making gossip about his sex life. Further complicating matters was the fact that Belinda was surely no bimbo. She ranked at number two on *Fortune* magazine's list of powerful businesswomen. And even after losing her bid to head up the Conservative Party in the spring of 2004, Belinda had, with Bill's backing, established herself as a force to be reckoned with in Canadian politics. Polls had shown, in fact, that she stood a chance of someday becoming Canada's first woman prime minister. In the meantime, Belinda's appointment to the cabinet or another high-level position in a Conservative government was almost assured. An affair with a lounge singer or an intern was one thing, but even rumors—however baseless—of a relationship between the former President and the beautiful Canadian politician would be impossible to ignore.

Nor, presumably, did Hillary appreciate the fact that Bill appeared to be serving as political mentor to another woman who wanted to run her country. Moreover, Stronach was a conservative who had more in common ideologically with George W. Bush than she did with the Clintons. "Bill Clinton must really like her a *lot,*" said one Canadian pol, "because they really don't have much in common when it comes to the issues. That must gall Hillary, I would think."

Keeping a wary eye north of the border, Hillary was now forced to cope with renewed pressure to endorse front-runner Dean. From her standpoint, a Dean candidacy was highly attractive; she could show her unflinching loyalty to the party by campaigning all-out for the ticket, never worrying that Dean might actually win and block her chances for a run in 2008.

But Hillary also remembered that, at this point in the nomination process back in 1991, polls showed her husband standing at only 4 percent among Democrats. As for who the Democratic candidate was going to be, Hillary allowed that it was "still a horse race." If she was to maintain her status as the most powerful figure in the party, she did not want to make the mistake of backing the wrong horse.

The senator's caution paid off on January 19, 2004, when Howard Dean, after losing to John Kerry in the Iowa caucuses, cut loose with a primal scream that effectively ended his presidential aspirations. Delivered on Martin Luther King Day, Dean's fatal remarks—ending in "Yeeeeeeeaarrrrrhhh!!"—would henceforth be known as his "I Have a Scream" speech.

"Oh, my *God*!" Hillary said when she first watched video of Dean's frenzied performance. She may have suspected that the scream would further undermine Dean's already slim chances of beating Bush. But at the time, Hillary apparently had no inkling that it would cost him the nomination.

Hillary watched with mounting concern as Dean's candidacy evaporated and John Kerry—the only candidate she rightly believed had any chance at all of defeating Bush in November—racked up one primary victory after another. She was equally wary of North Carolina's telegenic, charismatic John Edwards, who was now being touted as Kerry's probable vice-presidential pick. Behind closed doors, Hillary would oppose Edwards being offered a spot on the ticket. He did not, she argued, have sufficient foreign policy experience. More to the point, Edwards, a youthful-looking fifty, posed a very real threat to her future candidacy.

When it was clear that John Kerry had clinched the nomination, Hillary finally endorsed him on March 2—not in front of a bank of microphones on the Capitol steps, nor on the *Today* show or *Larry King Live,* but on Japanese television. "I will do everything I can," she told a reporter for the Nippon Television Network, "to get him elected." The nearly offhand remark was picked up by the wire services the next day.

Pressure for Hillary to join the Democratic ticket as Kerry's running mate was mounting by the minute. Were Kerry to beat Bush without Hillary on the ticket, she would "probably never be president of the United States," their old strategist Dick Morris observed. "This cold, hard fact is staring the Clintons in the face. . . ."

To be sure, Kerry would find himself in an awkward position if Hillary let it be known that she wanted the number two spot—a job that would leave her open to run in 2008 if Kerry lost and in 2012 if he won. The Clintons, who installed Terry McAuliffe as Democratic National Committee chairman, still controlled the party. When Kerry hinted that he intended to replace McAuliffe if he won the nomination, Hillary informed him in no uncertain terms that "Terry stays." Kerry backed down.

If Hillary wanted the second spot on the ticket, Kerry would have ample reason to give it to her. The Massachusetts senator, argued Morris, would be "pulling knives out of his back the entire race" if he "spurned" Hillary. "Kerry cannot afford to leave the Clintons sulking, like Achilles, in their tent. Otherwise, Troy will go Republican."

Yet Kerry, who worried that he would be overshadowed by Hillary if she ran with him, prayed that she wouldn't ask to be his running mate. "There are a lot of people—Democrats as well as Republicans—who really hate her," a Kerry aide said. "John knows that. They would turn out in the millions just to vote against her. And if John won, all the attention would be on her. *Nobody* in their right mind would want Hillary Clinton as their vice-president."

Underscoring that sentiment, Hillary was chosen as one of the "25 Toughest Guys in America" by *Men's Journal*—the first woman to make the list. Hillary offered a tongue-in-cheek objection to her twenty-fifth-place ranking behind rapper 50 Cent, who was shot nine times and drove himself to the hospital, and human crash-test dummy Rusty Haight, who volunteered to be in 740 car wrecks. Asked why Hillary was picked, the magazine's senior editor, Tom Foster, explained, "I think just looking at what she's been through and what she represents, that sort of stood for itself." Besides, Foster added, "would you mess with her?"

Sweeping into the Capital Hilton wearing a floor-length black satin coatdress that made her look like a distaff Darth Vader, Hillary took her place on the dais at Washington's annual Gridiron Club Dinner. Fittingly, she was on hand to trade one-liners with the man most likely to square off against her at the polls, either in her race for reelection to the Senate in 2006 or her own run for White House in 2008—Rudy Giuliani.

While W met with Mexican President Vicente Fox at the Bush ranch in Crawford, Texas, Hillary and Giuliani took good-natured jabs at each other. One of the former mayor's remarks was particularly revealing. "I fully suspect," Giuliani mused, "that come this November, when all is said and done, behind the sanctity of the voting booth curtain, I and Mrs. Clinton will be voting for the same person: George W. Bush."

Two days later, Hillary's longtime pal and staunch supporter Martha Stewart was found guilty on four counts of obstructing justice and lying to investigators regarding her fortuitously timed sale of ImClone stock—a conviction that could cost her $1 million in fines and land her behind bars for up to twenty years. She had raised hundreds of thousands of dollars for the Clintons, but now Martha was persona non grata in Hillaryland. Stewart had not heard a word from her dear friend the senator in the two years since her legal nightmare began. This was in stark contrast to other high-profile friends like Rosie O'Donnell and Bill Cosby, who showed up in court to lend their support.

Before the trial got under way, Martha had actually tried to contact the senator. According to friends of Stewart, Hillary refused to take Martha's calls. "She acted like she understood—she's a strong woman," said one. "But I think it broke her heart."

Still, by all accounts Martha appreciated the delicate position Hillary was in and was willing to overlook the fact that the senator appeared to have turned her back on a friend in need. Until, on the very day Martha was convicted, Senator Clinton delivered the coup de grâce. Once told of the verdict that would likely send her pal Martha to federal prison, Hillary took immediate action. "I'd better," she said without skipping a beat, "send her the money back"—the $1,000 donation, now presumably tainted by Stewart's criminal status, that Martha had made to Hillary's Senate campaign.

As her sentencing approached, Stewart sent a letter to one hundred friends asking them to write to Judge Miriam Goldman Cedarbaum and plead for leniency on her behalf. "If you would be so kind to write such a letter, please include your opinion of my character, my work ethic, my integrity and my probity," Martha suggested. "If possible, include any memorable experiences you have had with me to explain the basis of any expressed opinion(s)." Stewart then gave the address of the judge, as well as that of her legal team. Hillary received one of the letters, but made a conscious decision not to get involved.

Once Martha was out of jail, Hillary would have no trouble taking Stewart's money and employing the domestic diva's talents and connections to raise even more. But while Martha languished behind bars, the senator would have to keep her distance. So, too, would the former President. Hillary cautioned her husband not to contact their friend. For the moment, Hillary sighed, Martha's brush with the law had rendered her "radioactive."

In April 2004, the Clintons' strategy for putting Hillary in the White House took an intriguing turn when she told NBC's Katie Couric that she would not be John Kerry's running mate under any circumstances—even if he called and pleaded with her to take the number two spot on the Democratic ticket.

"I made it clear I don't want that to happen," insisted Hillary, who did the Couric interview to promote the paperback edition of her memoirs. "And what my answer will be—no—if it does happen. I'm not prepared to do that."

When Couric asked if she wanted to be President, Hillary replied, "It's not the way I think. I never thought I would end up being the senator from New York." Then, the woman who thirty years earlier had bragged to anyone who would listen that she was marrying a future President, told Couric, "I never thought that the long-haired, bearded boy I married in law school would end up being President. I don't think like that."

What Senator Clinton and the former President were really thinking became clearer just ten days later, when—as John Kerry began sinking in the polls—it was announced that Bill Clinton's memoirs would be published in late June. Despite pleas from the Kerry camp for Clinton to hold off on publication until after the general election in November, Bill timed the book's release and his subsequent publicity tour perfectly to steal the spotlight from John Kerry as he prepared for his coronation at the Democratic convention in late July. "The Clintons are obviously convinced Kerry can't win," said one veteran New York Democratic Party leader, "and they're doing what they can to make sure he doesn't. Hillary can talk about how much she wants Bush defeated, but with him in for another four years, she's got a clear shot at 2008. I mean, you've got to admire her chutzpah."

As Hillary once said of herself, "I don't quit. I keep going." Over the course of a thirty-five-year political career that stretched from the Black Panther and Vietnam antiwar movements to the impeachment of her scandal-plagued husband to her own Senate career, Hillary has never doubted for a moment that she should be the one in charge—of a student body, of a husband, of a nation. No investigation, no scandal, no charges of corruption, deception, conspiracy, perjury, or impropriety would impede her progress to power.

Ruthless. Brilliant. Grasping. Arrogant. Conniving. Compassionate. A victim. A saint. A schemer. All these words—and a few less savory ones—have been used to describe Hillary Rodham Clinton, just as they were used to describe Eva Perón. Like Evita, Hillary was the architect of her husband's rise to the presidency. Like Evita, Hillary was condemned by some and praised by others for wielding power as unofficial co-President. Like Evita, Hillary, along with her husband, was accused of corruption on a massive scale. Like Evita, Hillary dreamed of becoming her country's first woman President, and plotted relentlessly to make her dream a reality.

But then Hillary had one thing Evita didn't:

The Plan.

ACKNOWLEDGMENTS

For better or worse, Hillary Rodham Clinton is one of those larger-than-life figures for whom the oft-abused word *icon* was created. To millions, she is nothing short of heroic—Joan of Arc crossed with Eleanor Roosevelt. Equal numbers view her as the Antichrist. It is hard to believe that only now, after a dozen years spent fanning the flames of controversy, Hillary is at last embarking on a political career of her own—the second, highly anticipated (and to some, nervous-making) installment of The Plan.

Over the course of nine years and ten books, I have had the great pleasure of working with the talented people at William Morrow. Once again I am indebted to my editor, Maureen O'Brien, who brought the same high degree of commitment, insight, and editorial skill to *American Evita* that she did to *George and Laura: Portrait of an American Marriage* and *Sweet Caroline: Last Child of Camelot.* I am grateful, as ever, to all my friends at Morrow and HarperCollins, especially Jane Friedman, Cathy Hemming, Michael

Morrison, Laurie Rippon, Lisa Gallagher, Debbie Stier, Lindsey Moore, Michelle Corallo, James Fox, Mark Jackson, Beth Silfin, Chris Goff, Kyran Cassidy, Richard Aquan, Brad Foltz, Kim Lewis, Betty Lew, Christine Tanigawa, Jo Ann Metsch, and Camille McDuffie of Goldberg-McDuffie Communications.

After running out of ways last year to thank my longtime agent, Ellen Levine of the Trident Media Group, I resorted to expressing my gratitude in Japanese. Well, we've passed the twenty-two-year mark as agent and author, and I've come up dry again. So this time it's *muchas gracias,* Ellen, for your friendship above all else. My thanks as well to Ellen's talented associates, Melissa Flashman, Julia TerMatt, Sara Crowe, and Shannon Firth.

No son could have asked for wiser parents than Edward and Jeanette Andersen, though I'm certain they would be the first to take issue with that. Though they are two distinct personalities, my daughters Kate and Kelly share many common traits—brains, beauty, the courage of their convictions, and, most important, an abiding sense of humor about themselves and the world around them. They inherited all of these qualities from their mother, Valerie—the person to whom I owe the most thanks of all.

Additional thanks to Ed Koch, Judith Hope, Rick Lazio, John Peavoy, Marsha Laufer, Carol Yeldell Staley, Paul Fray, Alan Schecter, Steve Baranello, Juanita Broaddrick, Ernest Dumas, Richard Schaeffer, Susie Tompkins Buell, Dr. Arthur Curtis, Robert Moschetti, L. D. Brown, Kathleen Willey, Clare Ackerman, Dolly Kyle Browning, Dick Morris, Ann Henry, the late Daniel Patrick Moynihan, Cliff Jackson, Woody Bassett, John Perry Barlow, Esme Taylor, Bonnie Engle, David Leopoulos, Richard Winnie, Larry Nelson, Richard Atkinson, Rene Rockwell, Dr. Bobby Roberts, Morris Henry, the late Robert Treuhaft, Ellery Gordon, Joan Watkins, Rudy Moore, Parker Ladd, Paula Dranov, Frances Peavoy, Bernice Kizer, Joanne Webber, Joseph Ruggiero, Thomas Mars, Gaye Landau-Leonard, Adam Parkhomenko, Joe Dillard, Lawrence

C. Moss, Brian Farber, Debbie Larkin, Marsha Smolens, Madelaine Miller, Jill Iscoll, Simon Webber, Harry Criner, Robert McCleskey, Vada Sheid, Carl Whillock, Larry Gleghorn, John Kearney, Joe Newman, Jeanette Peterson, Jane Barber Smith, Stephen Pollack, Ernie Wright, Tom Freeman, Connie Burns, Brownie Ledbetter, Ed Coulter, Marcia Smolens, Brigit Dermott, Janet Lizop, Mary Sheid, Robert Lange, Sally Greenspan, Nancy Weiss, Dudley Freeman, Andrew Nurnberg, Anne Vanderhoop, Michelle Lapautre, Valerie Wimmer, Gary Gunderson, Bill Bushong, Kenneth P. Norwick, Larry Klayman, Lawrence R. Mulligan, David McGough, Lucianne Goldberg, Lou Ann Vogel, Larry Schwartz, Ray Whelan Sr., Hazel Southam, Rosemary McClure, Jean Chapin, Everett Raymond Kinstler, Barry Schenck, Tobias Markowitz, Wes Holmes, Yvette Reyes, Ray Whelan Jr., Betsy Loth, Arturo Santos, Betty Monkman, Ray Whelan, Kevin Hurley, Park Ridge Public Library, the Gunn Memorial Library, the White House Historical Association, the New York Public Library, the Clinton Presidential Center, the Clinton Materials Project, the George Bush Presidential Library and Museum, the Lyndon Baines Johnson Library, the Litchfield Library, the Federal Bureau of Investigation, the Bancroft Library at the University of California at Berkeley, the New Milford Library, the Southbury Public Library, Wellesley College, the Woodbury Library, the Silas Bronson Library, the Brookfield Library, the University of Arkansas, Oxford University, Yale University, Georgetown University, the Arkansas *Democrat-Gazette,* the Associated Press, Reuters, AP/Wide World, Globe Photos, Sipa, Corbis, and Design to Printing.

SOURCES AND CHAPTER NOTES

The following chapter notes have been compiled to give an overview of the sources drawn upon in writing *American Evita,* but they are by no means all-inclusive. The author has respected the wishes of many interviewed sources who wish to remain anonymous, and accordingly has not listed them either here or elsewhere in the text. These include officials still serving in Washington and in state and local government, as well as friends, schoolmates, neighbors, colleagues, advisers, fund-raisers, volunteers, mentors, protégés, volunteers, and staff members scattered across the globe. The archives and oral history collections of numerous institutions—including the Clinton Presidential Center, the John F. Kennedy Library, Wellesley College, Georgetown University, Yale University, Oxford University, Columbia University, and the University of Arkansas—yielded a wealth of information. Court documents and sworn depositions generated by numerous civil and criminal investigations—ranging from the testimony of the Arkansas State Troopers in the Paula Jones case to the mountain of testimony and evidence contained in the Starr Report to the investigations into Pardongate and 9/11—proved valuable. Needless to say, there have also been thousands of news reports and articles concerning Hillary Clinton that predate any coverage of her husband—all the way back to her controversial Wellesley address in 1969. Over the decades, these reports on Hillary appeared in such publications as the *New York Times,* the *Washington Post,* the *Wall Street Journal,* the *Boston Globe,* the *Arkansas Democrat,* the *Arkansas Gazette,* the *Chicago Tribune,* the *Los Angeles Times, Vanity Fair, The New Yorker, Time, Life,*

Newsweek, the *New York Observer, U.S. News & World Report, USA Today,* the (London) *Times, Paris-Match, Le Monde,* and carried on the wires of Reuters, Knight-Ridder, Gannett, and the Associated Press.

CHAPTERS 1 AND 2

Interview subjects included Ed Koch, Judith Hope, John Peavoy, Alan Schecter, Carolyn Yeldell Staley, Dolly Kyle Browning, Paul Fray, Dr. Arthur Curtis, David Leopoulos, Juanita Broaddrick, Robert Treuhaft, Woody Bassett, Joe Dillard, Robert Lange, Cliff Jackson, Ann Henry, Stephen Pollack, Betty Monkman. Published sources included Todd S. Purdum, "Striking Strengths, Glaring Failures," *New York Times,* December 24, 2000; Weston Kosova, "Backstage at the Finale," *Newsweek,* February 26, 2001; Margaret Carlson, "The Shadow Moves On," *Time,* January 29, 2001; Debra Rosenberg and Michael Isikoff, "Thinkin' About Tomorrow: Final Days," *Newsweek,* January 29, 2001; Kenneth Bazinet, "Prez Says Looong Goodbye," New York *Daily News,* January 21, 2001; Michael Medved, "When Bill Met Hillary," *Sunday Times* (London), August 21, 1994; Howard Fineman, "The Longest Goodbye," *Newsweek,* February 26, 2001; David Maraniss, *First in His Class* (New York: Simon & Schuster, 1995); David Freeman, "The First Friends," *Newsday,* January 6, 1993; Gail Sheehy, *Hillary's Choice* (New York: Ballantine, 1999); Martha Sherrill, "The Education of Hillary Clinton," *Washington Post,* January 11, 1993; Virginia Kelley, *Leading with My Heart* (New York: Simon & Schuster, 1994); Judith Warner, *Hillary Clinton: The Inside Story* (New York: Penguin Books, 1993); Jim Moore, *Clinton: Young Man in a Hurry* (Fort Worth: The Summit Group, 1992); Ernest Dumas, *The Clintons of Arkansas: An Introduction by Those Who Knew Them Best* (Fayetteville: University of Arkansas Press, 1993); Hillary Rodham Clinton, *Living History* (New York: Simon & Schuster, 1993); Meredith Oakley, *On the Make: The Rise of Bill Clinton* (Washington: Regnery Publishing, 1994); Michael Morrison, "Who Is Dan Lasater?," *Wall Street Journal,* August 7, 1995; Dolly Kyle Browning, *Purposes of the Heart* (Dallas: DOCC, 1997); Lloyd Grove, "Hillary Clinton, Trying to Have It All," *Washington Post,* March 10, 1992; Michael Weiskopf, "Inside Bill's Last Deal," *Time,* January 29, 2001.

CHAPTERS 3 AND 4

For these chapters, the author drew on conversations with Juanita Broaddrick, Ernest Dumas, L. D. Brown, Rudy Moore, Woody Bassett, Judith Hope, Ann

Henry, Paul Fray, Kathleen Willey, David Leopoulos, Vada Sheid, Thomas Mars, Brownie Ledbetter, James McDougal, Larry Gleghorn, Dolly Kyle Browning, Carl Whillock, Richard Atkinson, Morris Henry. Among the published sources consulted: David Maraniss and Susan Schmidt, "Hillary Clinton and the Whitewater Controversy: A Close-up; Her Public Record Suggests Conflicts with Self-Portrait of Naivete," *Washington Post,* June 2, 1996; George Stephanopoulos, *All Too Human* (New York: Little, Brown and Company, 1999); Charles Babcock and Sharon LaFraniere, "The Clintons' Finances: A Reflection of Their State's Power Structure," *Washington Post,* July 21, 1992; David Twersky, "A New Kind of Democrat?", *Commentary,* February 1993; Jane Gross, "Clinton's Lost Half-Brother: To Neighbors, He's Just Leon," *New York Times,* June 22, 1993; Joshua Muravchik, "Lament of a Clinton Supporter," *Commentary,* August 1993; Art Harris, "Gennifer Flowers," *Penthouse,* December 1992; David Brock, "Troopergate: Living with the Clintons," *The American Spectator,* January 1994; Ambrose Evans-Prichard, "Sally Perdue and Bill Clinton," *Sunday Telegraph* (London), January 23, 1994, and July 17, 1994; Alessandra Stanley, "A Softer Image for Hillary Clinton," *New York Times,* July 13, 1992; Michael Kelly, "Saint Hillary," *New York Times,* May 23, 1993; Hillary Rodham Clinton, *It Takes a Village, and Other Lessons Children Teach Us* (New York: Simon & Schuster, 1996); Peter J. Boyer, "Life After Vince," *The New Yorker,* September 11, 1995; Donnie Radcliffe, *Hillary Rodham Clinton: A First Lady for Our Time* (New York: Warner Books, 1993); Connie Bruck, "Hillary the Pol," *The New Yorker,* May 30, 1994; L.J. Davis, "The Name of the Rose: An Arkansas Thriller," *The New Republic,* April 4, 1994; Ann Devroy and Susan Schmidt, "The Mystery in Foster's Office: Following Suicide, What Drove Associates' Actions?" *Washington Post,* December 20, 1995; Bob Woodward, *The Agenda* (New York: Pocket Books, 1995); Gary Wills, "Inside Hillary's Head," *Washington Post,* January 21, 1996; David Brock, *The Seduction of Hillary Rodham* (New York: The Free Press, 1996); Roger Morris, *The Clintons and Their America* (New York: Henry Holt, 1996); "Turning Fifty: Hillary Confronts a Birthday and a Newly Empty Nest," *Time,* October 20, 1997; Dick Morris, *Behind the Oval Office* (New York: Random House, 1997); "Hillary's Role: How Much Clout?" *Newsweek,* February 15, 1993; Kenneth T. Walsh, "Portrait of a Marriage," *U.S. News & World Report,* August 31, 1998; "The Politics of Yuck," *Time,* September 14, 1998; "Clinton Admits Lewinsky Liaison to Jury," *New York Times,* August 18, 1998; William C. Rempel and David Willman, "Starr Looks for a Pattern in Job Offers by Clinton Camp," *Los Angeles Times,* February 9, 1998; Ann Douglas, "The Extraordinary Hillary Clinton," *Vogue,* December 1998; "The Fight of Their Lives," *Newsweek,* December 21, 1998; "The Starr Report: The Independent Counsel's Complete Report to Congress on the

Investigation of President Clinton," September 9, 1998; Dorothy Rabinowitz, "Juanita Broaddrick Meets the Press," *Wall Street Journal,* February 19, 1999; Howard Kurtz, *Spin Cycle* (New York: Simon & Schuster, 1998); Gail Sheehy, "What Hillary Wants," *Vanity Fair,* May 1992; Andrew Morton, *Monica's Story* (New York: St. Martin's Press, 1999); James Bennet, "The Next Clinton," *New York Times Magazine,* May 30, 1999; Douglas Waller, "Just One Day at a Time," *Time,* February 26, 2001; Raymond Hernandez, "Not the Mrs. Clinton Washington Thought It Knew," *New York Times,* January 24, 2002.

CHAPTERS 5 AND 6

Information for these chapters was based in part on conversations with Judith Hope, Ed Koch, Richard Schaeffer, Steve Baranello, Dick Morris, Rick Lazio, Marsha Laufer, Madelaine Miller, Ernest Dumas, Daniel Patrick Moynihan, Ernie Wright, John Perry Barlow, Susie Tompkins Buell, Esme Taylor, Robert Moschetti, Jane Barber Smith, John Kearney, Priscilla McMillan, Bonnie Engle, John Peavoy, Gaye Landau-Leonard, Marsha Smolens, Marta Sgubin, Larry Klayman, George Guldi, Connie Burns, Lou Ann Vogel, Lawrence C. Moss, Paula Dranov, Steve Levy, and Joseph Ruggiero. Published sources include: Joyce Milton, *The First Partner* (New York: William Morrow, 1999); Evan Thomas and Debra Rosenberg, "Hillary's Day in the Sun," *Newsweek,* March 1, 1999; Lucinda Franks, "The Intimate Hillary," *Talk,* September 1999; Debra Rosenberg and Gregory Beals, "Hillary Makes Her Move," *Newsweek,* December 6, 1999; Tom Junod, "Hillary Clinton," *Esquire,* October 1999; Elizabeth Kolbert, "Running on Empathy," *The New Yorker,* February 7, 2000; Michael Tomasky, "Hillary's Turn," *New York,* April 3, 2000; Meredith Berkman, "Hillary Now," *Ladies' Home Journal,* June 2000; Tish Durkin, "The Hidden Hillary," *The Observer,* July 17, 2000; Susan Estrich, "The Trouble with Hillary," *Harper's Bazaar,* August 2000; Michael Tomasky, "The Next Campaign," *New York,* November 20, 2000; Patrick S. Halley, *On the Road with Hillary* (New York: Viking, 2002); Maureen Orth, "The Face of Scandal," *Vanity Fair,* June 2001; Gail Sheehy, "Hillary's Solo Act," *Vanity Fair,* August 2001; Gregg Birnbaum, "Hillary: My Hands Are Clean," *New York Post,* January 30, 2001; "Hil Mum on Pardons," New York *Daily News,* January 26, 2001; Michael Duffy and Karen Tumulty, "Hillary Says She Knows Nothing About Her Brother's Dealings with Her Husband, but a New Investigation May Change That," *Time,* March 5, 2001; Susan Page, "Who Gets a Pardon? It Depends Who Asks," *USA Today,* March 20, 2001; Greg B. Smith, "Witness: Hillary Clinton Knew of Hasidics' Convictions, Unhappiness," Knight-Ridder/Tribune News

Service, April 5, 2001; Jennifer Senior, "Hill Climbing," *New York,* April 2, 2001; Jane Gross, "Through the Voters' Eyes, Senator Clinton's Missteps," *New York Times,* February 10, 2001; Susan Page and Mimi Hall, "Pardon Drama Casts Wide Net," *USA Today,* February 23, 2001; Robert Sullivan, "Hillary's Turn," *Vogue,* March 2001; Vincent Morris, "Hill's Office Hijack: Outrage over D.C. Space Grab," *New York Post,* June 27, 2001; Marjorie Williams, "Scenes from a Marriage," *Vanity Fair,* July 2001; Kenneth R. Bazinet, "Two Probes Cast Shadow over Hillary," New York *Daily News,* August 12, 2001; Nicholas Lemann, "The Hillary Perspective," *The New Yorker,* October 8, 2001; Margaret Carlson, "A Shower of Gifts for Hillary and Bill: Why Did the Clintons Troll for Freebies They Can Surely Afford?", *Time,* February 5, 2001; Raymond Hernandez, "Clintons Accused of Not Disclosing Gifts' True Value," *New York Times,* February 13, 2002; Jonathan Alter, "Citizen Clinton," *Newsweek,* April 8, 2002; Deborah Orin, "Hill's Dirty Money," *New York Post,* August 9, 2002; Shannon McCaffrey, "Sen. Clinton Will Donate Funds from ImClone CEO," Associated Press, August 10, 2002; "Clinton Wines and Dines UK Supermodel," *Sunday Times* (London), April 28, 2002; Tim Cooper, "A Hint of Saffron," *New York Observer,* May 5, 2002; Adam Nagourney and Raymond Hernandez, "For Hillary Clinton, a Dual Role as Star and as Subordinate," *New York Times,* October 22, 2002; "Poll: Hillary Clinton Top Democratic 2004 Choice," CNN, December 21, 2002; Joel Siegel, "Hillary Clinton Slams Bush Plan vs. Terror," Knight-Ridder/Tribune Business News, January 25, 2003; "Hillary Clinton Raises Her National Profile," Gannett News Service, January 15, 2003; Michael Wilson, "Senator Clinton Offers a Cure for Foot-in-Mouth Disease," *New York Times,* July 10, 2003; Barbara Olson, *The Final Days* (Regnery, 2001); Jim Dwyer, "Senator Clinton Says No to '04, but Playfully Hints at 'Yes,'" *New York Times,* September 9, 2003; Katharine Q. Selye, "Candidate Joined Crowd with Push from Clintons," *New York Times,* September 18, 2003; David T. Cook, "Hillary Clinton," *Christian Science Monitor,* September 25, 2003; Jeff Zeleny, "Some Democrats Continue Rush for Hillary Clinton to Enter Presidential Race," Knight-Ridder/Tribune News Service, September 25, 2003; Associated Press, "Hillary Clinton Joins U.S. Troops for Afghanistan Thanksgiving," November 27, 2003; David Rohde, "Hillary Clinton Dines with GIs in Afghanistan," *New York Times,* November 28, 2003; "Hillary Makes Iraq Visit Hot on the Heels of George W. Bush," BBC News, November 29, 2003; Melanie Thernstrom, "Hillary Clinton on Top of the World: Alpha Woman of the Year," *More,* April 2004; William Safire, "And the Kerry Envelope, Please," *New York Times,* April 17, 2004.

BIBLIOGRAPHY

Aldrich, Gary. *Unlimited Access.* Washington: Regnery Publishing, 1998.

Alinsky, Saul. *Rules for Radicals: A Practical Primer for Realistic Radicals.* New York: Random House, 1971.

Allen, Charles F., and Jonathan Portis. *The Comeback Kid: The Life and Career of Bill Clinton.* New York: Birch Lane Press, 1992.

Andersen, Christopher. *Jackie After Jack: Portrait of the Lady.* New York: William Morrow, 1998.

———. *Jack and Jackie: Portrait of an American Marriage.* New York: William Morrow, 1996.

Ashmore, Harry. *Arkansas: A Bicentennial History.* New York: W. W. Norton, 1978.

Blumenthal, Sidney. *The Clinton Wars.* New York: Farrar, Straus & Giroux, 2003.

Brock, David. *The Seduction of Hillary Rodham.* New York: The Free Press, 1996.

———. *Blinded by the Right.* New York: Crown, 2002.

Browning, Dolly Kyle. *Purposes of the Heart.* Dallas: Direct Outstanding Creations, 1997.

Carpozi, George, Jr. *Clinton Confidential: The Climb to Power.* Del Mar, Calif.: Emery Dalton Books, 1995.

Carville, James, and Mary Matalin. *All's Fair.* New York: Simon & Schuster, 1995.

Clarke, Richard A. *Against All Enemies: Inside America's War on Terror.* New York: Free Press, 2004.

Clinton, Hillary Rodham. *It Takes a Village, and Other Lessons Children Teach Us.* New York: Simon & Schuster, 1996.

———. *Dear Socks, Dear Buddy.* New York: Simon & Schuster, 1998.

———. *Living History.* New York: Simon & Schuster, 2003.

Cole, Steve. *Ghost Wars: The Secret History of the CIA, Afghanistan, and Bin Laden, from the Soviet Invasion to September 10, 2001.* New York: Penguin Press, 2004.

Conason, Joe. *The Hunting of the President.* New York: St. Martin's Press, 2000.

Drew, Elizabeth. *On the Edge: The Clinton Presidency.* New York: Simon & Schuster, 1994.

Dumas, Ernest. *The Clintons of Arkansas.* Little Rock: University of Arkansas Press, 1993.

Exner, Judith Campbell, as told to Ovid Demaris. *My Story.* New York: Grove Press, 1977.

Flinn, Susan K. *Speaking of Hillary: A Reader's Guide to the Most Controversial Woman in America.* Ashland, Ore: White Cloud Press, 2000.

Fulbright, J. William. *The Arrogance of Power.* New York: Random House, 1967.

Gallen, David. *Bill Clinton as They Know Him.* New York: Gallen Publishing, 1994.

Gitlin, Todd. *The Sixties: Years of Hope, Days of Rage.* New York: Bantam, 1987.

Halley, Patrick S. *On the Road with Hillary.* New York: Viking, 2002.

Hamilton, Nigel. *Bill Clinton: An American Journey.* Random House, 2003.

Isikoff, Michael. *Uncovering Clinton: A Reporter's Story.* New York: Crown, 1999.

Johnson, Rachel. *The Oxford Myth.* London: Weidenfeld Nicholson, 1988.

Kelley, Virginia. *Leading with My Heart.* New York: Simon & Schuster, 1994.

Levin, Robert E. *Clinton: The Inside Story.* New York: S.P.I. Books, 1992.

Limbacher, Carl. *Hillary's Scheme.* New York: Crown Forum, 2003.

Maraniss, David. *First in His Class.* New York: Simon & Schuster, 1995.

Milton, Joyce. *The First Partner.* New York: William Morrow, 1999.

Moore, Jim. *Clinton: Young Man in a Hurry.* Fort Worth: The Summit Group, 1992.

Morris, Dick. *Behind the Oval Office.* New York: Random House, 1997.

———. *Off with Their Heads.* New York: ReganBooks, 2003.

Morris, Roger. *Partners in Power: The Clintons and Their America.* New York: Henry Holt, 1996.

Morton, Andrew. *Monica's Story.* New York: St. Martin's Press, 1999.

Nelson, Rex. *The Hillary Factor.* New York: Gallen Publishing, 1993.

Oakley, Meredith L. *On the Make: The Rise of Bill Clinton.* Washington: Regnery Publishing, 1994.

Olson, Barbara. *Hell to Pay.* Washington: Regnery Publishing, 1994.

———. *The Final Days.* Washington: Regnery Publishing, 2001.

Osborne, Claire G. *The Unique Voice of Hillary Rodham Clinton.* New York: Avon Books, 1997.

Radcliffe, Donnie. *Hillary Rodham Clinton: A First Lady for Our Time.* New York: Warner Books, 1993.

Reeves, Richard. *President Kennedy: Profile of Power.* New York: Simon & Schuster, 1993.

Sheehy, Gail. *Hillary's Choice.* New York: Ballantine, 1999.

Starr, John Robert. *Yellow Dogs and Dark Horses.* Little Rock: August House, 1987.

Stephanopoulos, George. *All Too Human.* New York: Little, Brown and Company, 1999.

Stewart, James B. *Blood Sport: The President and His Adversaries.* New York: Simon & Schuster, 1996.

Walker, Martin. *The President We Deserve.* New York: Crown, 1996.

Warner, Judith. *Hillary Clinton: The Inside Story.* New York: Penguin Publishing, 1993.

Woodward, Bob. *The Agenda: Inside the Clinton White House.* New York: Pocket Books, 1994.

———. *Shadow: Five Presidents and the Legacy of Watergate.* New York: Simon & Schuster, 1999.

———. *Plan of Attack.* New York: Simon & Schuster, 2004.

INDEX